00106-15

Introduction to Basic Rigging

Overview

A common activity at nearly every construction site is the movement of material and equipment from one place to another using various types of lifting gear. The procedures involved in performing this task are known as rigging. Not every worker will participate in rigging operations, but nearly all will be exposed to it at one time or another. This module provides an overview of the various types of rigging equipment, common hitches used during a rigging operation, and the related Emergency Stop hand signal.

Module Six

Trainees with successful module completions may be eligible for credentialing through the NCCER Registry. To learn more, go to www.nccer.org or contact us at **1.888.622.3720**. Our website has information on the latest product releases and training, as well as online versions of our *Cornerstone* magazine and Pearson's product catalog.

Your feedback is welcome. You may email your comments to **curriculum@nccer.org**, send general comments and inquiries to **info@nccer.org**, or fill in the User Update form at the back of this module.

This information is general in nature and intended for training purposes only. Actual performance of activities described in this manual requires compliance with all applicable operating, service, maintenance, and safety procedures under the direction of qualified personnel. References in this manual to patented or proprietary devices do not constitute a recommendation of their use.

Copyright © 2015 by NCCER, Alachua, FL 32615, and published by Pearson Education, Inc., New York, NY 10013. All rights reserved. Printed in the United States of America. This publication is protected by Copyright, and permission should be obtained from NCCER prior to any prohibited reproduction, storage in a retrieval system, or transmission in any form or by any means, electronic, mechanical, photocopying, recording, or likewise. To obtain permission(s) to use material from this work, please submit a written request to NCCER Product Development, 13614 Progress Blvd., Alachua, FL 32615.

From *Core Curriculum, Trainee Guide*, Fifth Edition. NCCER.
Copyright © 2015 by NCCER. Published by Pearson Education. All rights reserved.

00106-15
INTRODUCTION TO BASIC RIGGING

Objective

When you have completed this module, you will be able to do the following:

1. Identify and describe various types of rigging slings, hardware, and equipment.
    a. Identify and describe various types of slings.
    b. Describe how to inspect various types of slings.
    c. Identify and describe how to inspect common rigging hardware.
    d. Identify and describe various types of hoists.
    e. Identify and describe basic rigging hitches and the related Emergency Stop hand signal.

Performance Task

Under the supervision of your instructor, you should be able to do the following:

1. Demonstrate the proper ASME Emergency Stop hand signal.

Trade Terms

Block and tackle	Master link	Sling reach
Bridle	One-rope lay	Splice
Bull ring	Plane	Strand
Competent person	Qualified person	Tag line
Core	Rated capacity	Tattle-tail
Hitch	Rejection criteria	Threaded shank
Hoist	Rigging hook	Unstranding
Lifting clamp	Shackle	Warning yarn
Load	Sheave	Wire rope
Load control	Sling	
Load stress	Sling legs	

Industry Recognized Credentials

If you are training through an NCCER-accredited sponsor, you may be eligible for credentials from NCCER's Registry. The ID number for this module is 00106-15. Note that this module may have been used in other NCCER curricula and may apply to other level completions. Contact NCCER's Registry at 888.622.3720 or go to **www.nccer.org** for more information.

Note

This module is an elective. It is not required for successful completion of the *Core Curriculum*.

Contents

Topics to be presented in this module include:

1.0.0 Basic Rigging Equipment 1
   1.1.0 Slings 2
      1.1.1 Sling Tagging Requirements 3
      1.1.2 Synthetic Slings 4
      1.1.3 Alloy Steel Chain Slings 8
      1.1.4 Wire Rope Slings 9
   1.2.0 Sling Inspection 10
      1.2.1 Synthetic Sling Inspection 11
      1.2.2 Alloy Steel Chain Sling Inspection 13
      1.2.3 Wire Rope Sling Inspection 13
   1.3.0 Rigging Hardware 16
      1.3.1 Shackles 16
      1.3.2 Eyebolts 17
      1.3.3 Lifting Clamps 18
      1.3.4 Rigging Hooks 19
   1.4.0 Hoists 20
      1.4.1 Operation of Chain Hoists 20
      1.4.2 Hoist Safety and Maintenance 22
   1.5.0 Hitches 23
      1.5.1 Vertical Hitch 24
      1.5.2 Choker Hitch 24
      1.5.3 Basket Hitch 27
      1.5.4 The Emergency Stop Signal 27

Figures

Figure 1   Overhead crane 2
Figure 2   Mobile cranes 3
Figure 3   Identification tag 3
Figure 4   Web slings provide surface protection 4
Figure 5   Synthetic web sling shaping 5
Figure 6   Protective pads 5
Figure 7   Synthetic web sling warning yarns 5
Figure 8   Synthetic web slings 6
Figure 9   Eye-and-eye synthetic web slings 6
Figure 10 Synthetic web sling hardware end fittings 6
Figure 11 Synthetic endless-strand jacketed sling 6
Figure 12 Twin-Path® sling 7
Figure 13 Twin-Path® sling makeup 8
Figure 14 Tattle-tails 8

Figure 15 Fiber-optic inspection cable 8
Figure 16 Markings on alloy steel chain slings 9
Figure 17 Chain slings 9
Figure 18 Three-leg chain bridle sling 9
Figure 19 Wire rope sling 10
Figure 20 Wire rope components 10
Figure 21 Wire rope supporting cores 11
Figure 22 Sling damage rejection criteria 12
Figure 23 Damage to chains 14
Figure 24 Common types of wire rope damage 15
Figure 25 One-rope lay 15
Figure 26 Shackles 16
Figure 27 Wide-body shackle 17
Figure 28 Synthetic web sling shackle 17
Figure 29 Don't over-tighten the shackle pin 17
Figure 30 Shackle correctly positioned 17
Figure 31 Eyebolt variations 18
Figure 32 Using lifting clamps 18
Figure 33 Basic nonlocking clamp 19
Figure 34 Lifting clamps 19
Figure 35 Rejection criteria for lifting clamps 20
Figure 36 Rigging hooks. 21
Figure 37 Common rigging hook defects 21
Figure 38 Block and tackle hoist system 22
Figure 39 Types of chain hoists 22
Figure 40 Chain hoist gear system 23
Figure 41 Hoist suspended from a trolley system 23
Figure 42 Use rachet level hoists for vertical lifting 23
Figure 43 Single vertical hitch 24
Figure 44 Bridle hitch 24
Figure 45 Multiple-leg bridle hitch 25
Figure 46 Choker hitches 25
Figure 47 Choker hitch constriction 26
Figure 48 Double choker hitches 27
Figure 49 Double-wrap choker hitches 28
Figure 50 Basket hitch 29
Figure 51 Emergency Stop hand signal 29

Section One

1.0.0 Basic Rigging Equipment

Objective

Identify and describe various types of rigging slings, hardware, and equipment.
a. Identify and describe various types of slings.
b. Describe how to inspect various types of slings.
c. Identify and describe how to inspect common rigging hardware.
d. Identify and describe various types of hoists.
e. Identify and describe basic rigging hitches and the related Emergency Stop hand signal.

Performance Task

1. Demonstrate the proper ASME Emergency Stop hand signal.

Trade Terms

Block and tackle: A simple rope-and-pulley system used to lift loads.

Bridle: A configuration using two or more slings to connect a load to a single hoist hook.

Bull ring: A single ring used to attach multiple slings to a hoist hook.

Competent person: An individual capable of identifying existing and predictable hazards in the surroundings and working conditions which are unsanitary, hazardous, or dangerous to employees; an individual who is authorized to take prompt corrective measures to eliminate these issues.

Core: Center support member of a wire rope around which the strands are laid.

Hitch: The rigging configuration by which a sling connects the load to the hoist hook. The three basic types of hitches are vertical, choker, and basket.

Hoist: A device that applies a mechanical force for lifting or lowering a load.

Lifting clamp: A device used to move loads such as steel plates or concrete panels without the use of slings.

Load control: The safe and efficient practice of load manipulation, using proper communication and handling techniques.

Load stress: The strain or tension applied on the rigging by the weight of the suspended load.

Master link: The main connection fitting for chain slings.

One-rope lay: The lengthwise distance it takes for one strand of a wire rope to make one complete turn around the core.

Plane: A surface in which a straight line joining two points lies wholly within that surface.

Qualified person: A person who, through the possession of a recognized degree, certificate, or professional standing, or one who has gained extensive knowledge, training, and experience, has successfully demonstrated his or her ability to solve problems relating to the subject matter, the work, or the project.

Rated capacity: The maximum load weight a sling or piece of hardware or equipment can hold or lift. Also referred to as the working load limit (WLL).

Rejection criteria: Standards, rules, or tests on which a decision can be based to remove an object or device from service because it is no longer safe.

Rigging hook: An item of rigging hardware used to attach a sling to a load.

Shackle: Coupling device used in an appropriate lifting apparatus to connect the rope to eye fittings, hooks, or other connectors.

Sheave: A grooved pulley-wheel for changing the direction of a rope's pull; often found on a crane.

Sling: Wire rope, alloy steel chain, metal mesh fabric, synthetic rope, synthetic webbing, or jacketed synthetic continuous loop fibers made into forms, with or without end fittings, used to handle loads.

Sling legs: The parts of the sling that reach from the attachment device around the object being lifted.

Sling reach: A measure taken from the master link of the sling, where it bears weight, to either the end fitting of the sling or the lowest point on the basket.

Splice: To join together.

Strand: A group of wires wound, or laid, around a center wire, or core. Strands are laid around a supporting core to form a rope.

Tag line: Rope that runs from the load to the ground. Riggers hold on to tag lines to keep a load from swinging or spinning during the lift.

Tattle-tail: Cord attached to the strands of an endless loop sling. It protrudes from the jacket. A tattle-tail is used to determine if an endless sling has been stretched or overloaded.

Threaded shank: A connecting end of a fastener, such as a bolt, with a series of spiral grooves cut into it. The grooves are designed to mate with grooves cut into another object in order to join them together.

00106-15 Introduction to Basic Rigging

Unstranding: Describes wire rope strands that have become untwisted. This weakens the rope and makes it easier to break.

Warning yarn: A component of the sling that shows the rigger whether the sling has suffered too much damage to be used.

Wire rope: A rope made from steel wires that are formed into strands and then laid around a supporting core to form a complete rope; sometimes called cable.

Rigging is the planned movement of material and equipment from one location to another, using slings, hoists, or other types of equipment. Some rigging operations use a loader to move materials around a job site. Other operations require cranes to lift such loads. Two common types of cranes are overhead cranes (*Figure 1*) and mobile cranes (*Figure 2*). As with other types of equipment, cranes are available in many different configurations and capacities.

Rigging operations can be extremely complicated and dangerous. Do not experiment with rigging operations, and never attempt a lift without the supervision of an officially recognized, qualified person. A lift may appear simple while it is in progress. That is because the people performing the lift know exactly what they are doing; there is no room for guesswork. No matter whether rigging operations involve simple or complicated equipment, only qualified persons may perform rigging operations without supervision.

> **WARNING!**
> Although this module provides instruction and information about common rigging equipment and hitch configurations, it does not provide any level of certification. Any questions about rigging procedures or equipment should be directed to an instructor or supervisor. Always refer to the manufacturer's instructions for any type of rigging equipment.

1.1.0 Slings

During a rigging operation, the load being lifted or moved must be connected to the apparatus, such as a crane, that will provide the power for movement. The connector—the link between the load and the apparatus—is often a sling made of synthetic, chain, or wire rope materials. This section focuses on three types of slings:

- Synthetic slings
- Alloy steel chain slings
- Wire rope slings

Figure 1 Overhead crane.

1.1.1 Sling Tagging Requirements

All slings are required to have identification tags (*Figure 3*). An identification tag must be securely attached to each sling and clearly marked with the information required for that type of sling. For all three types of slings, that information will include the manufacturer's name or trademark and the rated capacity of the type of hitch used with that sling. The rated capacity is the maximum load weight that the sling is designed to carry. Rated capacity is technically referred to as working load limit (WLL). The rated capacity, or WLL, of a sling must never be exceeded. Overloading a sling can result in catastrophic failure.

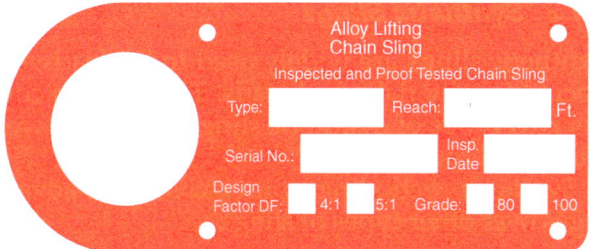

Figure 3 Identification tag.

WARNING! All slings and hardware have a rated capacity, or working load limit (WLL). Rated capacity, or WLL, is defined as the maximum load weight that a sling or piece of hardware or equipment can hold or lift. Under no circumstances should the rated capacity ever be exceeded. Overloading a piece of rigging equipment can result in catastrophic failure.

The following are the tagging requirements for synthetic slings:

- Manufacturer's name or trademark
- Manufacturer's code or stock number (unique for each sling)
- Rated capacities for the types of hitches used
- Type of synthetic material used in the manufacture of the sling

The following are the tagging requirements for alloy steel chain slings:

- Manufacturer's name or trademark
- Manufactured grade of steel
- Link size (diameter)
- Rated load and the angle on which the rating is based
- Sling reach
- Number of sling legs

The following are the tagging requirements for wire rope slings:

- Manufacturer's name or trademark
- Rated capacity in a vertical hitch (other hitches optional)
- Diameter of wire rope (optional)
- Manufacturer's code or stock number of sling (optional)

Any sling without an identification tag must be removed from service immediately, since its characteristics and rating can no longer be identified. A qualified person may be able to install a new tag in some cases.

Figure 2 Mobile cranes.

 00106-15   Introduction to Basic Rigging   Module Six   3

1.1.2 Synthetic Slings

Synthetic slings are widely used to lift loads, especially easily damaged loads. This section covers two types of synthetic slings: synthetic web slings and round slings.

Synthetic web slings provide several advantages over other types of slings:

- They are softer and wider than chain or wire rope slings. Therefore, they do not scratch or damage machined or delicate surfaces (*Figure 4*).
- They do not rust or corrode and therefore will not stain the loads they are lifting.
- They are lightweight, making them easier to handle than wire rope or chain slings. Most synthetic slings weigh less than half as much as a wire rope that has the same rated capacity. Some new synthetic fiber slings weigh one-tenth as much as wire rope.
- They are flexible. They mold themselves to the shape of the load (*Figure 5*).
- They are very elastic, and they stretch under a load much more than wire rope. This stretching allows synthetic slings to absorb shocks and to cushion the load.
- Loads suspended in synthetic web slings are less likely to twist than those in wire rope or chain slings.

For all the advantages that synthetic web slings provide, there are some concerns that must be kept in mind when using them. For example, synthetic web slings should not be exposed to temperatures above 180°F (82°C). They are also susceptible to cuts, abrasions, and other wear-and-tear damage. To prevent damage to synthetic web slings, riggers use protective pads (*Figure 6*).

> **CAUTION**
> If the sling does not come with protective pads, use other kinds of softeners of sufficient strength or thickness to protect the sling where it makes contact with the load. Pieces of old sling, fire hose, canvas, or rubber can be used. Manufactured softeners can also be purchased.

Most synthetic web slings are manufactured with red-core **warning yarns**. These are used to let the rigger know whether the sling has suffered too much wear or damage to be used. When the red yarns are exposed, the synthetic web sling should not be used (*Figure 7*).

Figure 4 Web slings provide surface protection.

Figure 5 Synthetic web sling shaping.

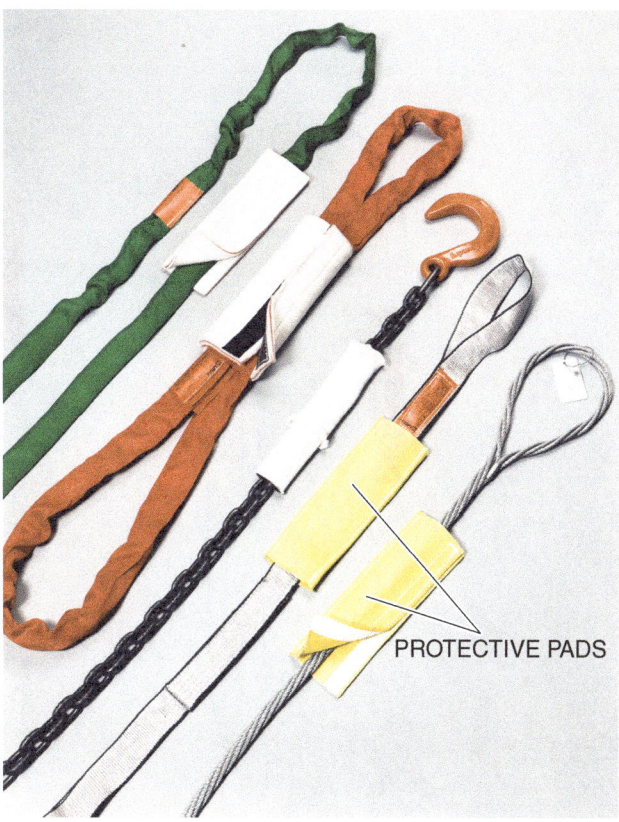

Figure 6 Protective pads.

Synthetic web slings are available in several designs. The most common are the following (*Figure 8*):

- Endless web slings, which are also called grommet slings.
- Synthetic web eye-and-eye slings, which are made by sewing an end of the sling directly to the sling body. Standard eye-and-eye slings (*Figure 9*) have eyes on the same plane as the sling material; twisted eye-and-eye slings have eyes at right angles to the main portion of the sling and are primarily used for choker hitches (which are discussed later).
- Round slings, which are endless and made in a continuous circle out of polyester filament yarn. The yarn is then covered by a woven sleeve.

Synthetic web eye-and-eye slings are also available with hardware end fittings instead of fabric eyes. The standard end fittings are made of either aluminum or steel. They come in male and female configurations (*Figure 10*).

Round slings are made by wrapping a synthetic yarn around a set of spindles to form an endless loop. A protective jacket encases the core yarn (*Figure 11*). One type of round sling is a Twin-Path® sling. Twin-Path® is a trademarked sling made by Slingmax®, Inc., although the term is sometimes used generically.

Twin-Path® slings are made of a synthetic fiber, such as polyester. The material used to make up the strands and the number of wraps in a loop determines the rated capacity of the sling, or how much weight it can handle. Aramid round slings, for example, have a greater rated capacity for their size than the web-type slings do.

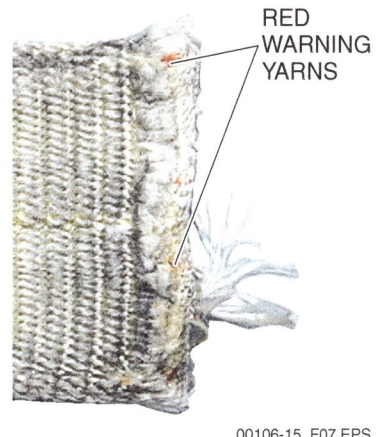

Figure 7 Synthetic web sling warning yarns.

00106-15 Introduction to Basic Rigging Module Six 5

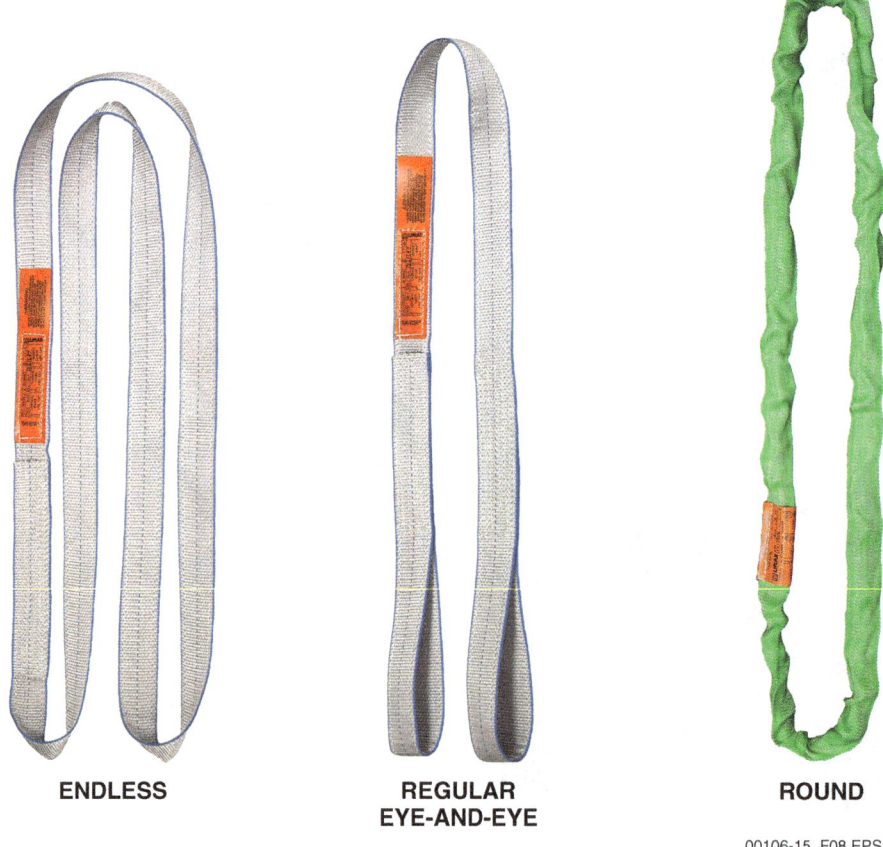

Figure 8 Synthetic web slings.

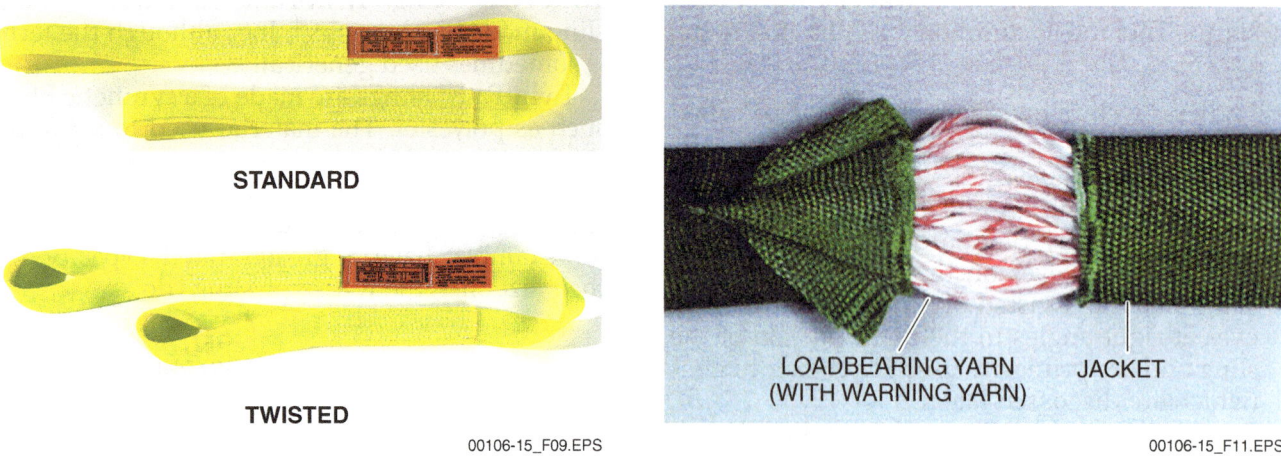

Figure 9 Eye-and-eye synthetic web slings.

Figure 11 Synthetic endless-strand jacketed sling.

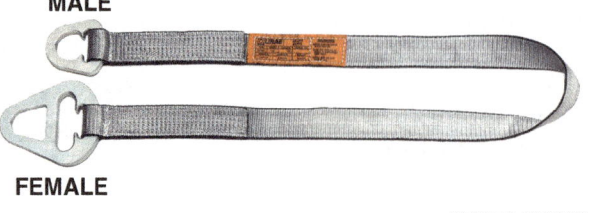

Figure 10 Synthetic web sling hardware end fittings.

Twin-Path® slings are also available in a design with two separate wound loops of strand jacketed together side-by-side (*Figure 12*). This design greatly increases the lifting capacity of the sling.

The jackets of these slings are available in several materials for various purposes, including heat-resistant Nomex®, polyester, and bulked nylon (Covermax™) (*Figure 13*). Twin-Path® slings featuring K-Spec® yarn weigh at least 50 percent less than a polyester round sling of the same size and capacity.

Twin-Path® slings are equipped with tattle-tail yarns to help riggers determine whether the sling has become overloaded or stretched beyond a safe limit (*Figure 14*). Note that these devices are different from red-core yarn, which is incorporated directly into the fabric of the sling. These slings are also available with a fiber-optic inspection cable running through the strand (*Figure 15*).

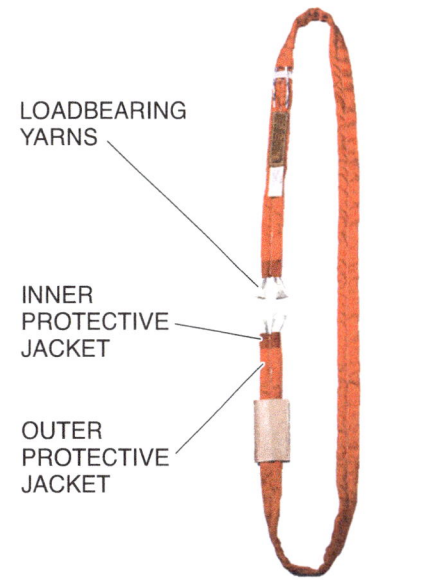

Figure 12 Twin-Path® sling.

Around the World
Heavy Lifters

The world's largest cranes are capable of lifting astonishing amounts of weight. Some of these megacranes are permanently stationary, some can be assembled and disassembled onsite, and some are mobile. Taisun, a fixed dual-beam gantry crane in China, can lift over 22,000 tons (over 20,000 metric tons). Mammoet, a manufacturer of heavy lifting equipment in the Netherlands, has built numerous cranes that are referred to as PTCs (Platform Twinring Containerised). These cranes can be disassembled and packed into shipping containers, unloaded at any dock in the world, and hauled by truck to a job site. This type of crane, as shown here, is assembled onto a stationary ring that enables the crane and counterweights to move in a 360-degree radius. These giant cranes can lift up to 3,200 tons (about 2,900 metric tons).

One of the largest mobile cranes is Liebherr's LR 13000 crawler crane. It has a more conventional design and appearance. Its maximum lifting capacity is over 3,300 tons (about 3,000 metric tons).

00106-15 Introduction to Basic Rigging

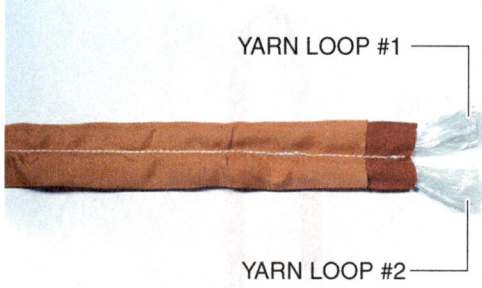

Figure 13 Twin-Path® sling makeup.

When a light is directed at one end of the fiber, the other end will light up to show that the strand has not been broken.

The way the sling will be used determines what material is used for the jackets of round slings. Polyester is normally used for light- to medium-duty sling jackets. Covermax™, a high-strength material with much greater resistance to cutting and abrasion, is used to make a sturdier jacket for heavy-duty uses.

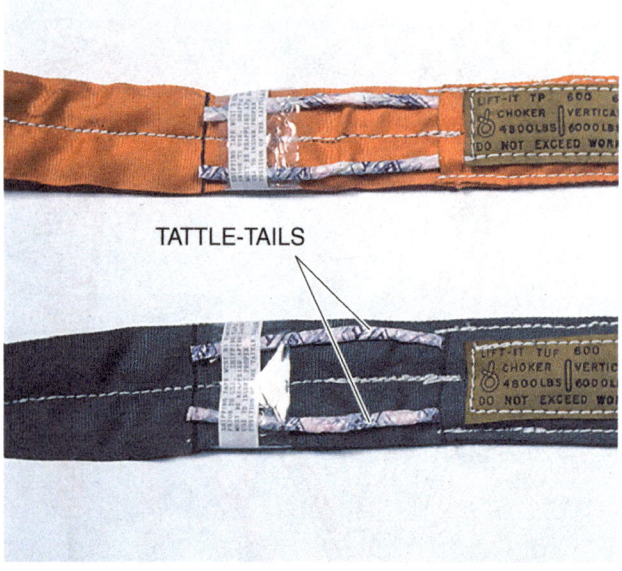

Figure 14 Tattle-tails.

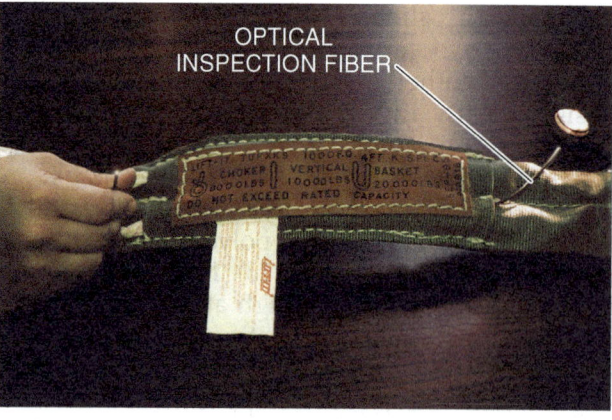

Figure 15 Fiber-optic inspection cable.

1.1.3 Alloy Steel Chain Slings

Alloy steel chain, like wire rope (which is covered later), can be used in many different rigging operations. Chain slings are often used for lifts in high heat or rugged conditions. Chain slings can be adjusted over a center of gravity, making it a versatile sling. They are also the most durable. However, a chain sling weighs much more than a wire rope sling, and it also may be harder to inspect. Workers will encounter both types of slings in the field and must decide which type of sling is best for each situation.

Steel chain slings used for overhead lifting must be made of alloy steel. The higher the alloy grade, the safer and more durable the chain is. Alloy steel chains commonly used for most overhead lifting are marked with the number 8, the number 80, the number 800, or the letter A (see *Figure 16*).

Steel chain slings have two basic designs with many variations:

- Single- and double-basket slings (*Figure 17*) do not require end-fitting hardware. The chain is attached to the master link in a permanent basket hitch or hitches.
- Chain bridle slings are available with two to four legs (*Figure 18*).

Regulations and Site Procedures

The information in this module is intended as a general guide. The techniques shown here are not the only methods that can be used to perform a lift. Many techniques can be safely used to rig and lift different loads.

Some of the techniques for certain kinds of rigging and lifting are spelled out in requirements issued by federal government agencies. Some will be provided at the job site, where written site procedures that address any special conditions that affect lifting procedures on that site can be found. Questions about any of these procedures should be directed to the supervisor at the site.

NCCER – *Core Curriculum* 00106-15

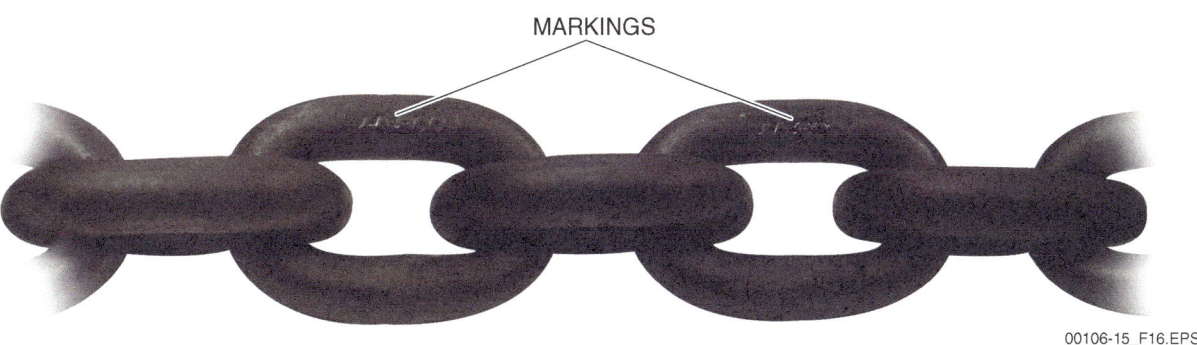

Figure 16 Markings on alloy steel chain slings.

> **CAUTION**
> Never drag alloy steel chain slings across hard surfaces, especially concrete. Friction with abrasive surfaces wears out the chain, weakens the sling, and often damages identifying markings.

Alloy steel chain slings are used for lifts when temperatures are high or where the slings will be subjected to steady and severe abuse. Most alloy steel chain slings can be used in temperatures up to 500°F (260°C) with little loss in rated capacity.

Even though alloy steel chain slings can withstand extreme temperature ranges and abusive working conditions, these slings can be damaged if loads are dropped on them; if they are wrapped around loads with sharp corners (unless softeners are used); or if they are exposed to intense temperatures.

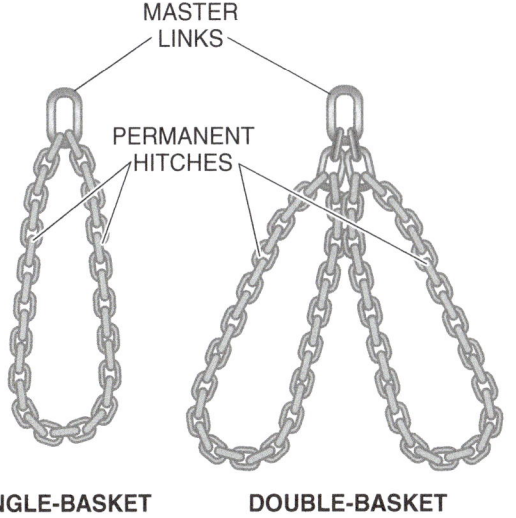

Figure 17 Chain slings.

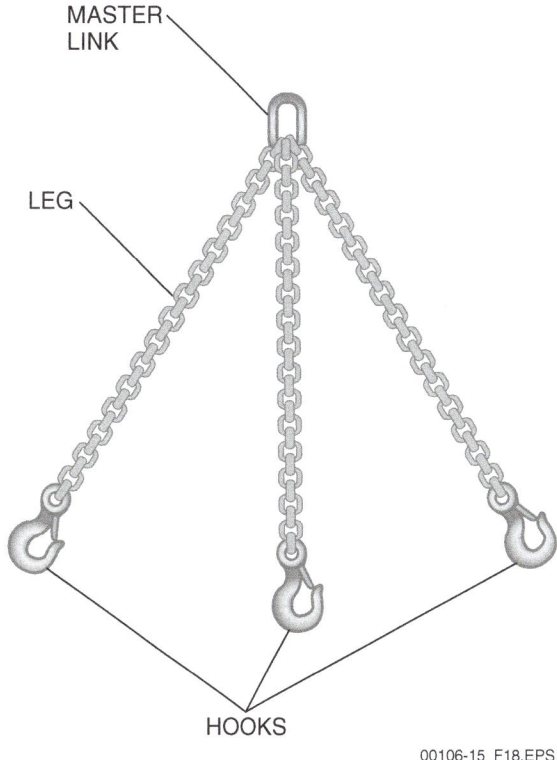

Figure 18 Three-leg chain bridle sling.

1.1.4 Wire Rope Slings

Wire rope slings (*Figure 19*) are made of high-strength steel wires that are formed into strands and wrapped around a supporting core. While there are many types of wire rope, they all share this common design. As a general rule, wire rope slings are lighter than chain, can withstand substantial abuse, and are easier to handle than chain slings. They can also withstand relatively high temperatures.

The wire rope that makes up a wire rope sling is designed so that the strand wires and the supporting core interact with one another by sliding and adjusting. This movement compensates for the ever-changing stresses placed upon the rope.

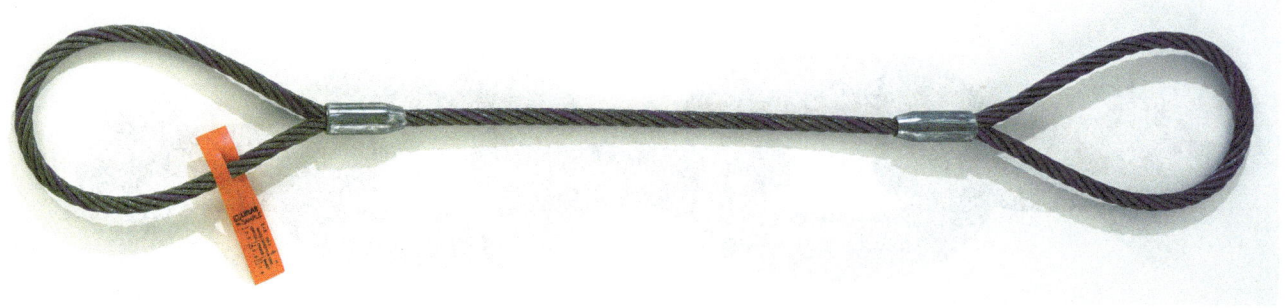

Figure 19 Wire rope sling.

It makes it less likely that the wires and the core will be damaged when the rope is bent around **sheaves** or loads, or when it is placed at an angle during rigging. The wires must be able to move the way they were designed to move, so a wire rope's rated capacity depends on it being in good condition.

Figure 20 shows the three basic components of a wire rope: the supporting core, high-grade steel wires, and multiple center wires.

The supporting core enables the strands to keep their original shapes. There are three basic types of supporting cores for wire rope (*Figure 21*):

- *Fiber cores* – Usually made of synthetic fibers, but can also be made from natural vegetable fibers, such as sisal.
- *Independent wire rope cores* – Made of a separate wire rope with its own core and strands; the core rope wires are much smaller and more delicate than the strand wires in the outer rope.
- *Strand cores* – Made by using one strand of the same size and type as the rest of the strands of rope.

The materials used for the supporting core have both desirable characteristics and drawbacks, depending on how they are to be used. Fiber core ropes, for example, may be damaged by heat at relatively low temperatures (180°F to 200°F [82°C to 93°C]), as well as by exposure to caustic chemicals.

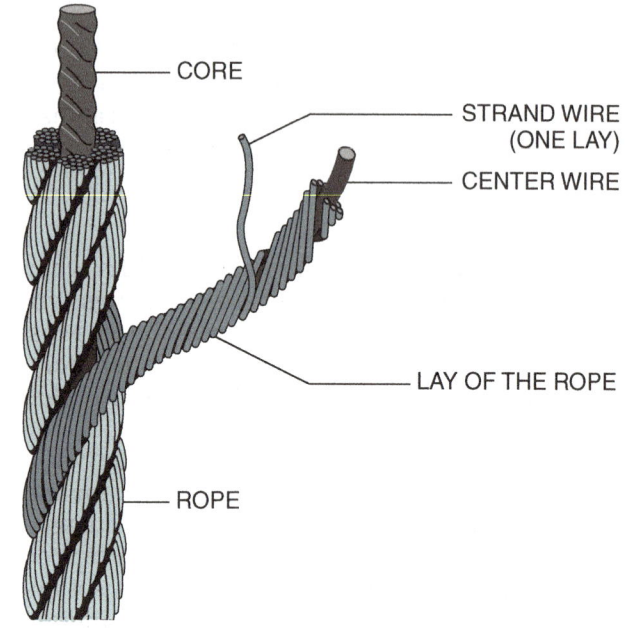

Figure 20 Wire rope components.

1.2.0 Sling Inspection

Rigging operations can be dangerous even in the best conditions. If equipment failure occurs, the results can be devastating. Since slings are commonly used to connect loads to lifting equipment, they are susceptible to wear and damage. To ensure that slings are safe to use, it is absolutely critical that they be thoroughly inspected on a regular basis. Inspections must be conducted regardless of any existing markings or tags that are related to previous inspections.

> **WARNING!**
> Although this module provides instruction and information about lifting component inspection, it does not provide any level of certification. Any questions about rigging procedures or equipment should be directed to an instructor or supervisor. Always refer to the manufacturer's instructions for any type of rigging equipment.

> **Did You Know?**
> **Sling Cut Protection**
> Cuts are the primary cause of synthetic sling failure. Often these cuts come from making contact with sharp edges on the load. Besides sling jackets, or wear pads, devices called edge protectors are often placed on the edges of a load to help reduce the likelihood of cutting the sling.

10   NCCER – *Core Curriculum*   00106-15

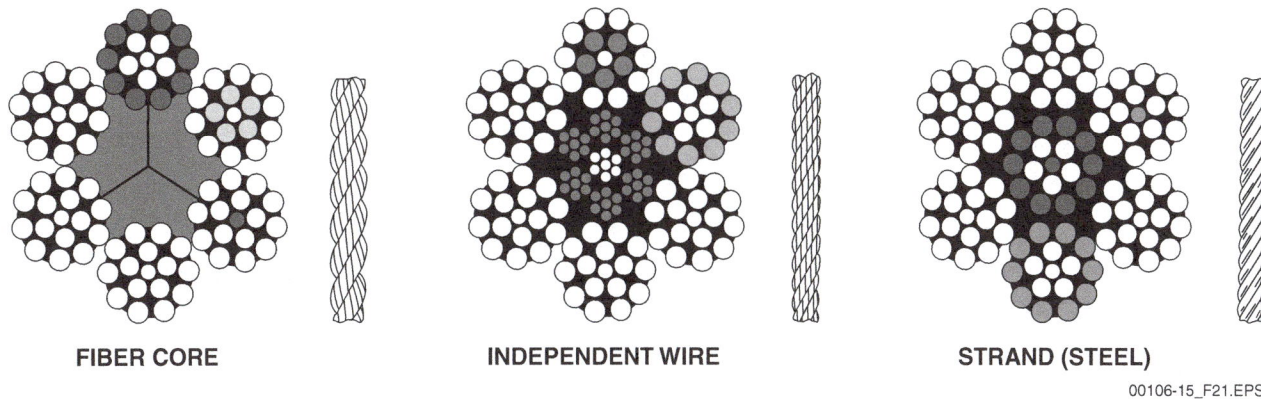

FIBER CORE **INDEPENDENT WIRE** **STRAND (STEEL)**

Figure 21 Wire rope supporting cores.

1.2.1 Synthetic Sling Inspection

Like all slings, synthetic slings must be inspected before each use to determine whether they are in good condition and can be used. They must be inspected along the entire length of the sling, both visually (looking at them) and manually (feeling them). If any rejection criteria are met, the sling must be removed from service.

> **WARNING!**
> A competent person must inspect synthetic slings before they are used, every time. The inspection must involve a visual and a touch examination. Sometimes defects and damage can be seen; other times they can be felt.

If any synthetic sling meets any of the rejection criteria presented in this section, it must be removed from service immediately. In addition, the rigger has to exercise sound judgment. Along with looking for any single major problem, the rigger must also watch for combinations of relatively minor defects in the sling. Combinations of minor damage may make the sling unsafe to use, even though the specific defects found may not be listed in the rejection criteria.

Workers who help inspect synthetic slings must alert a qualified person if any defects are suspected, especially if any of the following synthetic sling damage rejection criteria are found (*Figure 22*):

- A missing identification tag or a tag that cannot be read. Any synthetic sling without an identification tag must be removed from service immediately.
- Abrasion that has worn through the outer jacket or has exposed the loadbearing yarn of the sling, or abrasion that has exposed the warning yarn of a web sling.
- A cut that has severed the outer jacket or exposed the loadbearing yarn (single-layer jacket) of a round sling, or has exposed the warning yarn of a web sling.
- A tear that has exposed the inner jacket or the loadbearing yarns (single-layer jacket) of a round sling, or has exposed the warning yarn of a web sling.
- A puncture.
- Broken or worn stitching in the splice or stitching of a web sling.
- A knot that cannot be removed by hand in either a web or round sling.
- A snag in the sling that reveals the warning yarn of a web sling or tears through the outer jacket, or exposes the loadbearing yarns of a round sling.
- Crushing of either a web sling or a round sling. Crushing in a web sling feels like a hollow pocket or depression in the sling. Crushing in a round sling feels like a hard, flat spot underneath the jacket.
- Damage from overload (often called tensile damage or overstretching) in a web or round sling. Tensile damage in a web sling is evident when the weave pattern of the fabric begins to pull apart. Twin-Path® slings have tattle-tail features. When the tail has been pulled into the jacket, it indicates that the sling may have been overloaded.
- Chemical damage, including discoloration, burns, and melting of the fabric or jacket.
- Heat damage, ranging from friction burns to melting of the sling material, the loadbearing strands, or the jacket. Friction burns give the webbing material a crusty or slick texture. Heat damage to the jacket of a round sling looks like glazing or charring. In round slings with heat-resistant jackets, you may not be able to see heat damage to the outside of the sling; however, the internal yarns may have been damaged. You can detect that damage by carefully handling the sling and feeling for brittle or fused fibers inside the sling, or by flexing and folding the sling and listening for the sound of fused yarn fibers cracking or breaking inside.

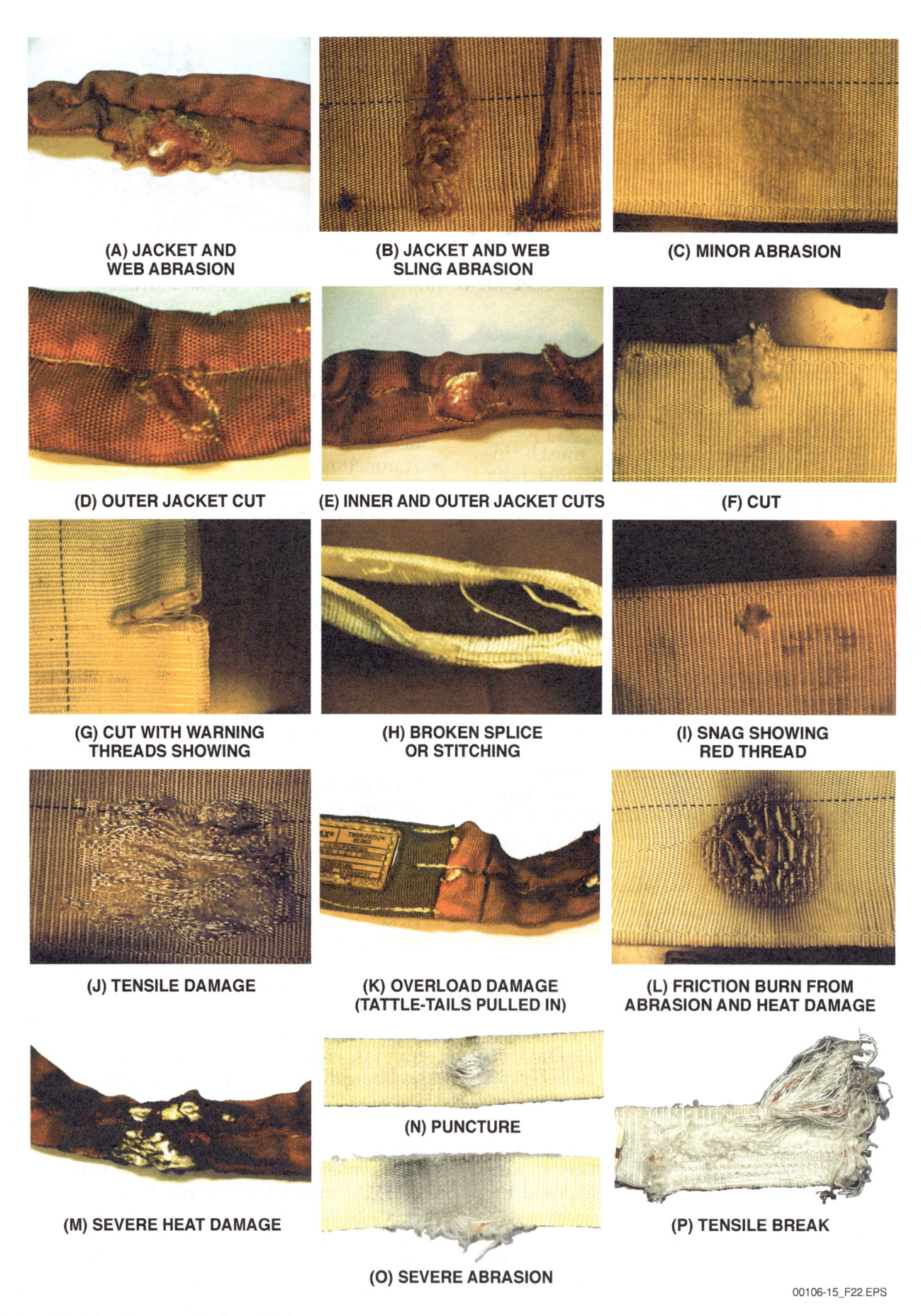

Figure 22 Sling damage rejection criteria.

- Ultraviolet (UV) damage. The evidence of UV damage is a bleaching-out of the sling material, which breaks down the synthetic fibers. Web slings with UV damage will release a powder-like substance when they are flexed and folded. Round slings, especially those made of Cordura® nylon and other specially treated synthetic fibers, have a much greater resistance to UV damage. In round slings, UV damage shows up as a roughening of the fabric texture where no other sign of damage, such as abrasion, can be found.
- Loss of flexibility caused by the presence of dirt or other abrasives. A sling that has lost its flexibility becomes stiff. Abrasive particles embedded in the sling material act like tiny blades that cut apart the internal fibers of the sling every time it is stretched, flexed, or wrapped around a load.

1.2.2 Alloy Steel Chain Sling Inspection

Alloy steel chain slings must be carefully inspected before each use to determine if they are safe to use. There are numerous rejection criteria that indicate when an alloy steel chain sling must be removed from service. A chain sling may also need to be removed from service if something shows up that does not exactly match the rejection criteria. Not all riggers are qualified to make decisions about the condition of the equipment being used. If questions arise about whether a sling is defective or unsafe, a qualified person should be consulted. Note that the chain links in a chain sling cannot be repaired by replacement.

An alloy steel chain sling must be removed from service for any of the following defects (*Figure 23*):

- Missing or illegible identification tag
- Cracks
- Heat damage
- Stretched links; damage is evident when the link grows long and when the barrels—the long sides of the links—start to close up
- Bent links
- Twisted links
- Evidence of replacement links that have been used to repair the chain
- Excessive rust or corrosion, meaning rust or corrosion that cannot be easily removed with a wire brush
- Cuts, chips, or gouges resulting from impact on the chain
- Damaged end fittings, such as hooks, clamps, and other hardware
- Excessive wear at the link-bearing surfaces
- Scraping or abrasion

1.2.3 Wire Rope Sling Inspection

Like other slings, wire rope slings must be inspected before each use by a competent person. Remember that a competent person is defined by OSHA as a person who is capable of identifying existing and predictable hazards in the surroundings or working conditions that are unsanitary, hazardous, or dangerous to employees, and who has authorization to take prompt corrective measures to eliminate those conditions.

If the wire rope is damaged, it must be removed from service. Only a competent person can decide whether to use a wire rope in a rigging operation or discard it if it is damaged. Any worker who suspects that a wire rope sling may be damaged must bring it to the attention of a competent person, stopping an active lift if necessary to do so. If a wire rope sling is simply missing its tag or the tag has become unreadable due to wear, it can be retagged by a qualified person. Remember that a qualified person is defined by OSHA as a person who, by possession of a recognized degree, certificate, or professional standing, or by extensive knowledge, training, and experience, has demonstrated the ability to solve or prevent problems relating to a certain subject, work, or project.

Sling Jackets

If only the outer jacket of a sling is damaged, under the rejection criteria, that sling can be sent back to the manufacturer for a new jacket. It must be tested and certified by the manufacturer before it is used again. It is much less costly to do this than to replace the sling. However, slings that are removed because of heat- or tension-related jacket damage cannot be returned for repair and must be disposed of by a qualified person.

Following an inspection, a wire rope sling may be rejected based on several common types of damage (*Figure 24*), including broken wires, kinks, birdcaging, crushing, corrosion and rust, and heat damage.

- *Broken wires* – Broken wires in the strands of a wire rope weaken the material strength of the rope and interfere with the interaction among the rope's moving parts. External broken wires usually mean normal fatigue, but internal or severe external breaks should be investigated closely. Internal or severe breaks in a wire rope mean it has been used improperly. Rejection criteria for broken wires consider how many wires are broken in one lay length of rope, or **one-rope lay**. One-rope lay (identifying a single rope lay and not a rope lay made with one rope) is a term that defines the lengthwise distance it takes for one strand of wire to make one complete turn around the core (*Figure 25*). Different wire ropes have different one-rope lays, so it is important for a competent person to inspect each wire rope closely when looking for broken wires.

WARNING! Broken wires in a sling are extremely sharp and can easily cut or puncture the skin. Never use a bare hand to handle wire rope slings or to inspect them by running a hand along their length.

HEAT DAMAGE AND CRACK

IMPACT DAMAGE BENT LINKS

Link barrel has bent from being wrapped around a load with sharp corners.

Link bent from impact.

EXCESSIVE WEAR

Links wear at the bearing surfaces.

OVERLOAD DAMAGE

As the link stretches the barrels will close up.

TWISTED LINKS

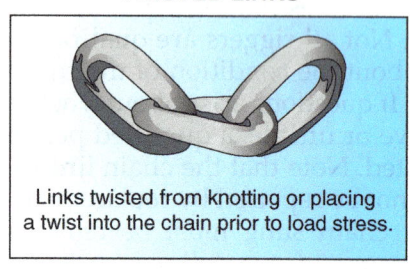

Links twisted from knotting or placing a twist into the chain prior to load stress.

CUTS, CHIPS, AND GOUGES

RUST AND CORROSION

Figure 23 Damage to chains.

Figure 24 Common types of wire rope damage.

Figure 25 One-rope lay.

for the change in the stress level by adjusting inside the strands. The built-up stress then finds its own release out through the strands. Birdcaging usually occurs in an area where already-existing damage prevents the wires from moving to compensate for changes in stress, position, and bending of the rope. Any sign of birdcaging is cause to remove the rope from service immediately.

- *Unstranding* – Unstranding describes wire rope strands that have become untwisted. This weakens the rope and makes it easier to break.
- *Signs of core failure* – Failure or breakage within the core is usually spotted by the protrusion of core strands through or into the outer jacket of strands.
- *Crushing* – This results from setting a load down on a sling or from hammering or pounding a sling into place. Crushing of the sling prevents the wires from adjusting to changes in stress, changes in position, and bends. A crushed sling usually results in the crushing or breaking of the core wires directly beneath the damaged strands. If crushing occurs, the sling must be removed from service immediately.
- *Corrosion and rust* – Corrosion and rust of wire rope are the result of improper or insufficient lubrication. Corrosion and rust are considered excessive if there is surface scaling or rust that cannot easily be removed with a wire brush, or if they occur inside the rope. If corrosion and rust are excessive, the rope must be removed from service.
- *Heat damage* – Heat damage makes wires in the strands and core in the affected area become brittle. Heat damage appears as discoloration and sometimes the actual melting of the wire rope. A wire rope that has been damaged by heat must be removed from service.
- *Integrity of end connections* – Any end connections that have been applied to the wire rope must be inspected for any signs of damage or failure, such as marks indicating the connection has slipped or moved.

- *Kinks* – Kinking, or distortion of the rope, is a very common type of damage. Kinking can result in serious accidents. Sharp kinks restrict or prevent the movement of wires in the strands at the area of the kink. This means the rope is damaged and must not be used. Ropes with kinks in the form of large, gradual loops in a corkscrew configuration must be removed from service.
- *Birdcaging* – This damage occurs when a load is released too quickly and the strands are pulled or bounced away from the supporting core. The wires in the strands cannot compensate

1.3.0 Rigging Hardware

Rigging hardware is as crucial as the crane, the slings, or any specially designed lifting frame or hoisting device. If the hardware that connects the slings to either the load or the master link were to fail, the load would drop just as it would if the crane, hoist, or slings were to fail. Hardware failure related to improper attachment, selection, or inspection contributes to a great number of the deaths, serious injuries, and property damage events in rigging accidents. The importance of hardware selection, maintenance, inspection, and proper use cannot be stressed enough. The requirements for rigging hardware are as stringent as those governing cranes and slings.

1.3.1 Shackles

A *shackle* is an item of rigging hardware used to attach an item to a load or to couple slings together. For example, a shackle can be used to couple the end of a wire rope to eye fittings or hooks. It consists of a U-shaped body and a removable pin.

Shackles used for overhead lifting should be made from forged steel, not cast steel. Quenched and tempered steel is the preferred material because of its increased toughness, but at a minimum, shackles must be made of drop- or hammer-forged steel.

All shackles must have a stamp that is clearly visible, showing the manufacturer's trademark, the size of the shackle (determined by the diameter of the shackle's body, not by the diameter of the pin), and the rated capacity of the shackle.

Shackles are available in two basic classes, identified by their shapes: anchor shackles and chain shackles. Both anchor and chain shackles have three basic types of pin designs, each one unique, as shown in *Figure 26*. The screw pin shackle design is the most widely used type in general industry. Shackle pins can be threaded into the shackle (screw pin); be threaded and use a nut to secure them; use a locking pin through the end; or be both threaded and pinned to secure them to the shackle.

Specialty shackles are available for specific applications where a standard shackle would not work well. For example, wide-body shackles (*Figure 27*) are for heavy-lifting applications.

SCREW PIN ANCHOR SHACKLE

SCREW PIN CHAIN SHACKLE

ROUND PIN ANCHOR SHACKLE

ROUND PIN CHAIN SHACKLE

SAFETY ANCHOR SHACKLE

SAFETY CHAIN SHACKLE

Figure 26 Shackles.

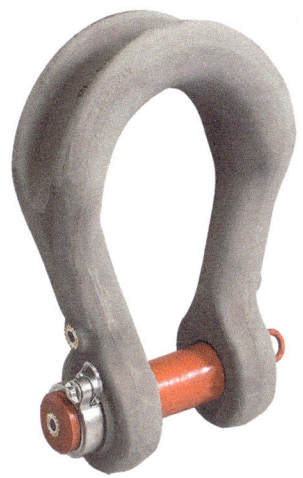

Figure 27 Wide-body shackle.

Synthetic web sling shackles (*Figure 28*) are designed with a wide throat opening and a wide bow that is contoured to provide a larger, nonslip surface area to accommodate the wider body of synthetic web slings.

When a screw pin shackle is used, it is important to tighten the pin the proper amount. The pin should be screwed in until it is fully engaged and the shoulder of the pin is in contact with the shackle body (*Figure 29*). If the pin is left loose, the shackle can stretch under load. If the pin is over-tightened, the load stress from the sling can torque the pin even more. This can stretch and jam the pin's threads.

When any type of shackle is used in a rigging operation, it must be positioned so that the load is centered in the bow of the shackle and not on the shackle pin (*Figure 30*). If the shackle uses a screw pin and the load is positioned on the pin, the shifting of the load could cause the pin to twist and unscrew.

Shackles, like any other type of hardware, must be inspected by the rigger before each use to make sure there are no defects that would make the shackle unsafe. Each lift may cause some degree of damage or may further reveal existing damage.

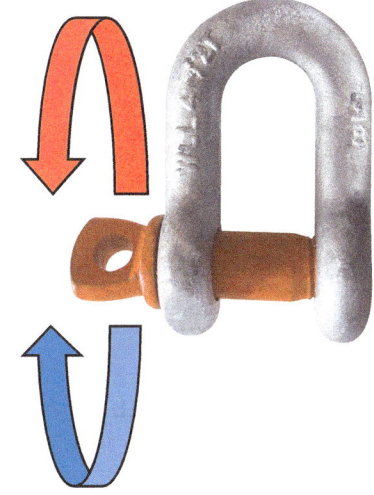

Figure 29 Don't over-tighten the shackle pin.

If any of the following conditions exists, a shackle must be removed from use:

- Bends, cracks, or other damage to the shackle body
- Incorrect shackle pin or improperly substituted pin
- Bent, broken, or loose shackle pin
- Damaged threads on threaded shackle pin
- Missing or illegible capacity and size markings

1.3.2 Eyebolts

An eyebolt is an item of rigging hardware with a threaded shank. The eyebolt's shank end is attached directly to the load, and the eyebolt's eye end is used to attach a sling to the load.

Eyebolts for overhead lifting should be made of drop- or hammer-forged steel. Eyebolts are available in three basic designs with several variations (*Figure 31*).

Figure 28 Synthetic web sling shackle.

Figure 30 Shackle correctly positioned.

00106-15   Introduction to Basic Rigging   Module Six   17

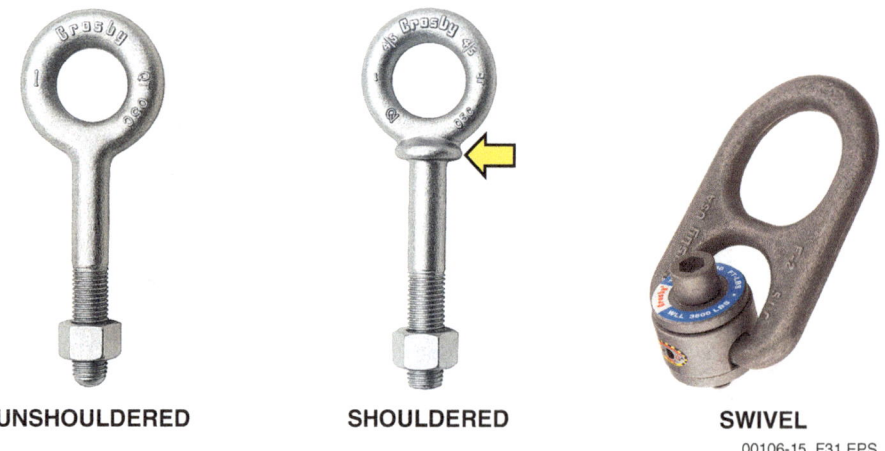

Figure 31 Eyebolt variations.

Unshouldered eyebolts are designed for straight vertical pulls only. Shouldered eyebolts have a shoulder that is used to help support the eyebolt during pulls that are slightly angular. Swivel eyebolts, also called hoist rings, are designed for angular pulls from 0 to 90 degrees from the horizontal plane of the load.

In a few cases, eyebolts must be torqued to a specific value when they are installed for lifting use.

1.3.3 Lifting Clamps

Lifting clamps are used to move loads such as steel plates, sheet piles, large pipe, or concrete panels without the use of slings. These rigging devices are designed to bite down on a load and use the jaw tension to secure the load (*Figure 32*).

All lifting clamps must be made of forged steel, and they must be stamped with their rated capacity. Some clamps, such as the one in *Figure 33*, use the weight of the load to produce and sustain the clamping pressure; the grip tightens as the load increases. Others use an adjustable cam that is set and tightened to maintain a secure grip on the load.

Lifting clamps are designed to carry one item at a time, regardless of the capacity or jaw dimensions of the clamp or the thickness or weight of the item being lifted. In order for the clamp to hold an item securely, the cam and the jaw must bite or grip both sides of a single item. Placing more than one plate or sheet into the clamp prevents both the cam and the jaw from securing both sides of the plate or sheet. The clamps must be placed to ensure that the load remains balanced. For larger loads, two lifting clamps are typically required.

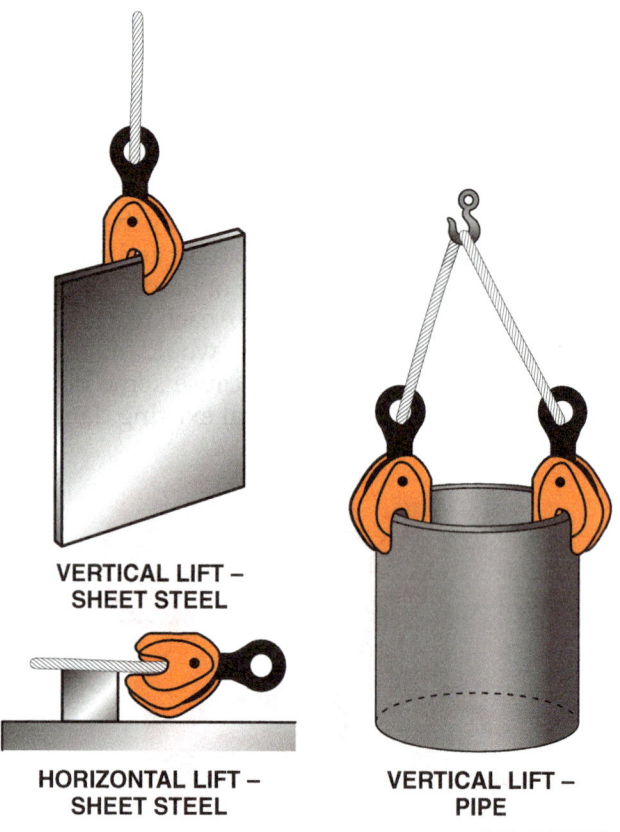

Figure 32 Using lifting clamps.

> **Did You Know?**
> ### Shackles and Pins
> Pins should never be swapped in shackles. The thread forms of different brands of shackles and different brands of pins may not engage properly with one another. In addition, there is no easy way to tell how much reserve strength is left in a pin or a shackle. Further, the rated capacities of different shackles and pins may not be compatible. The shackle pin should be placed and secured into a single shackle between uses.

Figure 33 Basic nonlocking clamp.

Lifting clamps are available in a wide variety of designs, so it is important to match the type of clamp with the intended application. *Figure 34* shows some lifting clamps designed for specific uses.

Lifting clamps, like any other type of rigging hardware, must be inspected by the rigger before every use to make sure there are no defects that would make the clamp unsafe. Each lift may cause some degree of damage or may further reveal existing damage.

If any of the following conditions exists, a clamp must be removed from use (*Figure 35*):

- Cracks
- Abrasion, wear, or scraping
- Any deformation or other impact damage to the shape that is detectable during a visual examination
- Excessive rust or corrosion, meaning rust or corrosion that cannot be removed easily with a wire brush
- Excessive wear of the teeth
- Heat damage
- Loose or damaged screws or rivets
- Worn springs

1.3.4 Rigging Hooks

A **rigging hook** is an item of rigging hardware used to attach a sling to a load. There are many classes of rigging hooks used, but most rigging hooks fall into a few basic categories (*Figure 36*). Most rigging hooks have safety latches or gates to prevent slings or other connectors from accidentally coming out of the hook during use. Hooks that do not have safety latches are usually designed so that a latch can be added.

- Eye hooks are the most common type of end fitting hook. A rigging eye hook has a large eye that can accommodate large couplers.
- Reverse eye hooks position the point of the hook perpendicular to the eye.
- Sliding choker hooks are installed onto the sling when it is made. The hooks, which can be positioned anywhere along the sling body, are used to secure the sling eye in a choker hitch. Sliding choker hooks are available for steel chain slings.
- Grab hooks are used on steel chain slings. These hooks fit securely in the chain link, so that choker hitches can be made and chains can be shortened.
- Shortening clutches, a more efficient version of the grab hooks, provide a secure grab of the shortened sling leg with no reduction in the capacity of the chain because the clutch fully supports the links.

LOCKING CLAMP

SCREW-ADJUSTED CLAMP

NON-MARRING

Figure 34 Lifting clamps.

00106-15   Introduction to Basic Rigging   Module Six   19

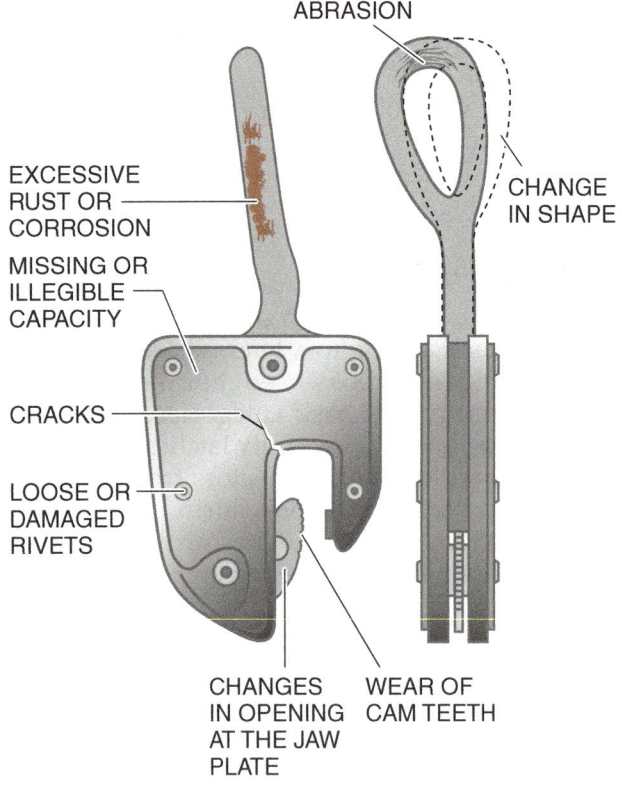

Figure 35 Rejection criteria for lifting clamps.

Hooks used for rigging must be made of drop- or hammer-forged steel. Although most rigging hooks have safety latches or gates, some hooks used for special applications may not have them. If a safety latch is installed in a rigging hook, the latch must be in good working condition. Damaged safety latches can be easily replaced. Any damage to a safety latch must be reported to an instructor or supervisor.

When hooks are installed as end fittings, they must be inspected along with the rest of the sling before each use. Slings with hook-type end fittings need to be removed from service for any of the following defects (*Figure 37*):

- Wear, scraping, or abrasion
- Cracks
- Cuts, gouges, nicks, or chips
- Excessive rust or corrosion, meaning rust or corrosion that cannot be easily removed with a wire brush
- An increase in the throat opening of the hook—easy to detect if the hook is equipped with a safety latch, because the latch will no longer bridge the throat opening
- A twist in the hook
- An elongation of the hook
- A broken or missing safety latch

1.4.0 Hoists

A hoist is a device that uses pulleys or gears to provide a mechanical advantage for lifting a load, allowing objects to be lifted that cannot be lifted manually. Some hoists are mounted on trolleys and use electricity or compressed air for power. In this section, a simple hoisting mechanism called a **block and tackle**, and a more complex hoisting mechanism called a chain hoist, (*Figure 38*) are discussed.

A block and tackle is a simple rope-and-pulley system used to lift loads. By using fixed pulleys and a wire rope attached to a load, a rigger can raise and lower the load by pulling the rope or by using a winch to lift the load. A block and tackle assembly multiplies the power applied, so that a worker can lift a much heavier load than could be lifted without it.

Chain hoists may be operated manually or mechanically. There are three types of chain hoists (*Figure 39*): manual, electric, and pneumatic. Because electric and pneumatic chain hoists use mechanical power, they are known as powered chain hoists. All chain hoists use a gear system to lift heavy loads (*Figure 40*). The gearing is coupled to a sprocket that has a chain with a hook attached to it. The load is hooked onto a chain and the gearing turns the sprocket, causing the chain to travel over the sprocket and move the load. The hoist can be suspended by a hook connected to an appropriate anchorage point, or it can be suspended from a trolley system (*Figure 41*).

1.4.1 Operation of Chain Hoists

Chain hoists are operated by hand or by electric or pneumatic power. Some of the fundamental operating procedures for hand chain hoists and powered chain hoists are as follows:

- *Hand chain hoist* – To use a hand chain hoist, the rigger suspends the hoist above the load to be lifted, using either the suspension hook or the trolley mount. The rigger then attaches the hook to the load and pulls the hand chain drop to raise the load. The load will either rise or fall, depending on which side of the chain drop is pulled.
- *Powered chain hoist* – To use an electric or a pneumatic powered chain hoist, the rigger positions the chain hoist on the trolley above the load to be lifted, attaches the hook to the load, and uses the control pad to operate the hoist. Only qualified persons may use powered chain hoists.

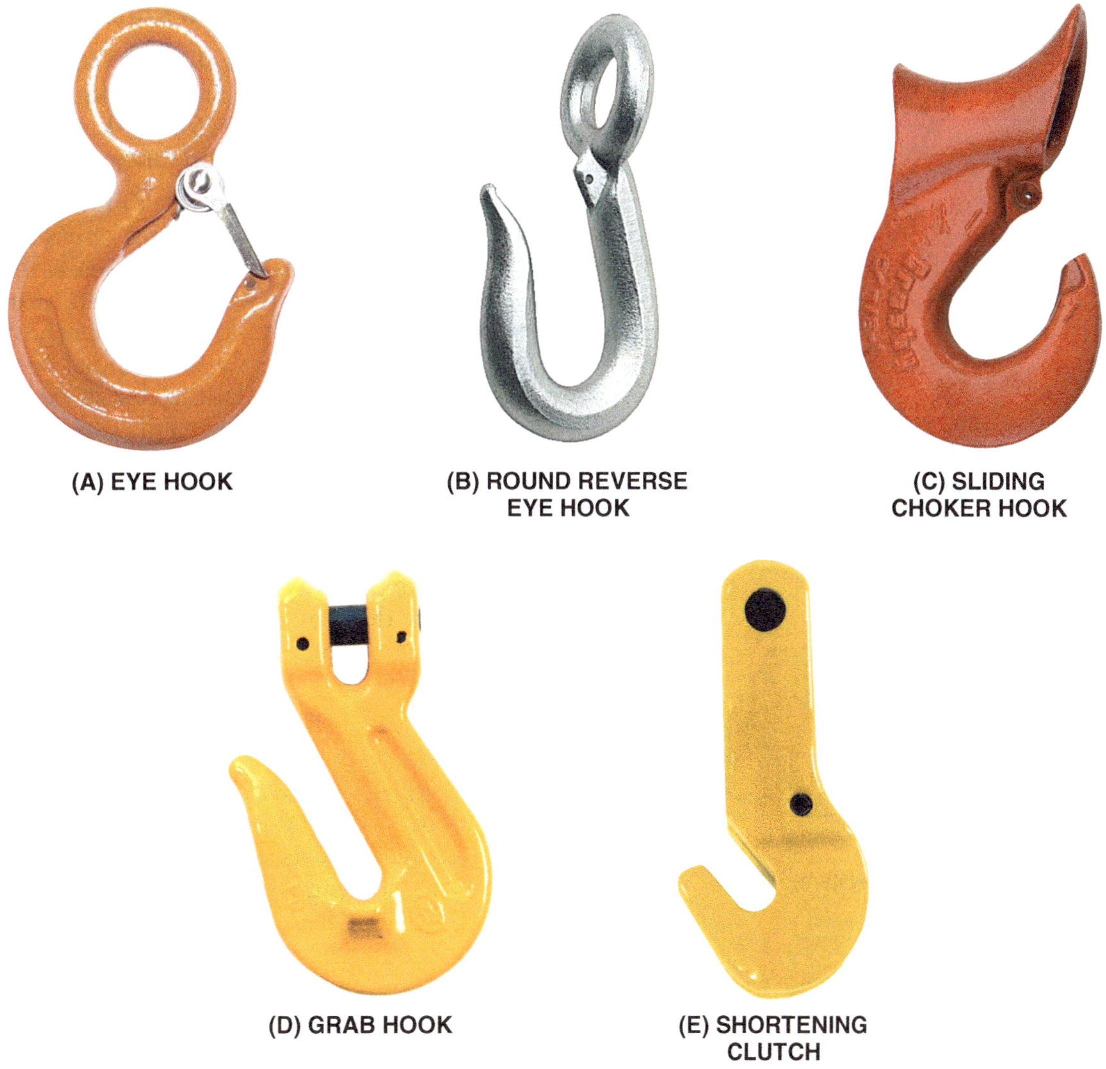

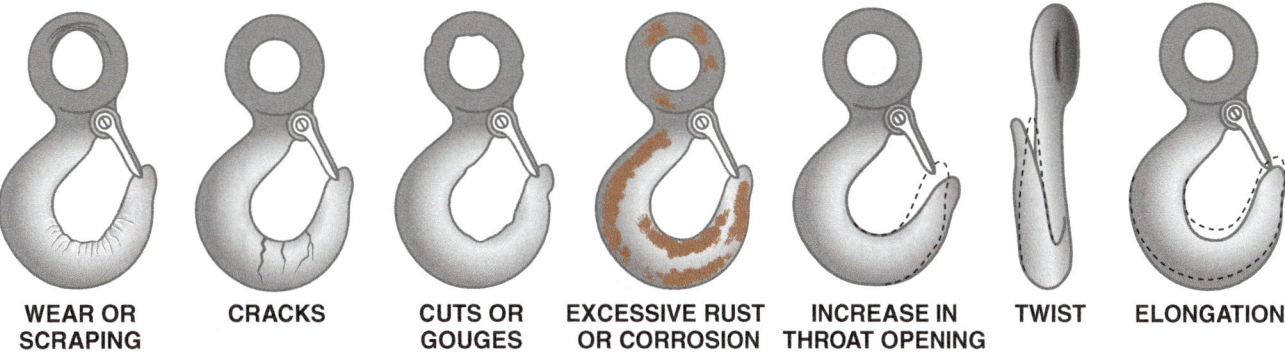

(A) EYE HOOK

(B) ROUND REVERSE EYE HOOK

(C) SLIDING CHOKER HOOK

(D) GRAB HOOK

(E) SHORTENING CLUTCH

Figure 36 Rigging hooks.

WEAR OR SCRAPING

CRACKS

CUTS OR GOUGES

EXCESSIVE RUST OR CORROSION

INCREASE IN THROAT OPENING

TWIST

ELONGATION

Figure 37 Common rigging hook defects.

00106-15 Introduction to Basic Rigging Module Six 21

Figure 38 Block and tackle hoist system.

1.4.2 Hoist Safety and Maintenance

In addition to the general safety rules that were presented in the *Basic Safety* module, there are some specific safety rules for working with hoists. Observe the following guidelines:

- Always use the appropriate PPE when working with and around any lifting operations. This includes a hard hat, safety glasses, safety shoes, and a high-visibility vest (at a minimum).
- Make sure that the load is properly balanced and attached correctly to the hoist before a lift is attempted. Unbalanced loads can slide or shift, causing the hoist to fail.
- Keep gears, chains, and ropes clean. Improper maintenance can shorten the working life of chains and ropes.
- Lubricate gears periodically to keep the wheels from freezing up. All such maintenance work must be done by a competent person.
- Never perform a lift of any size without proper supervision.

Never use a common come-along for vertical overhead lifting. Use a come-along only to move loads horizontally over the ground. Be careful not to confuse a come-along with a ratchet lever hoist. Ratchet lever hoists (*Figure 42*) have both a friction-type holding brake and a ratchet-and-pawl

ELECTRIC

PNEUMATIC

MANUAL

Figure 39 Types of chain hoists.

22   NCCER – *Core Curriculum*   00106-15

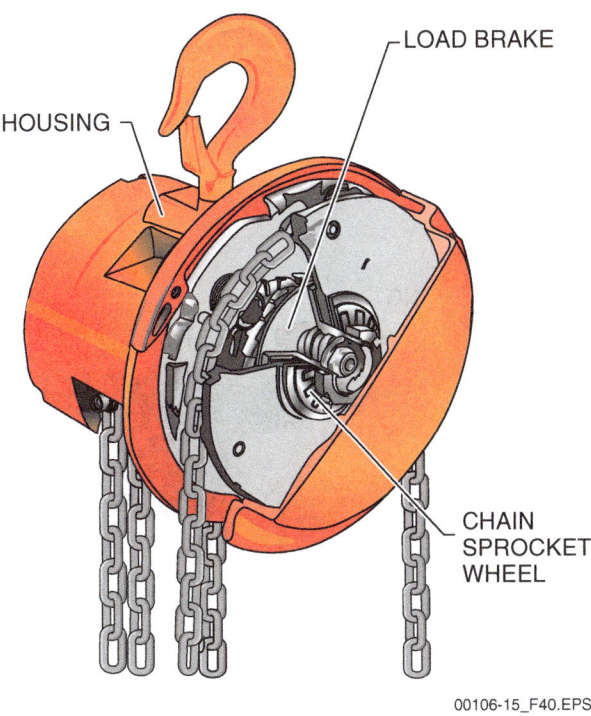

Figure 40 Chain hoist gear system.

load control brake. Cable come-alongs have only a spring-load ratchet that holds the pawl in place to secure the cable. If the ratchet and pawl fail, an overhead load falls. Be certain that any device used to lift is designed and specified for that task. Among manufacturers, the terms *come-along* and *ratchet lever hoist* are often used in the description of a single item, creating confusion in the marketplace and on the job. In most cases, cable come-alongs are not designed for lifting loads.

1.5.0 Hitches

In a rigging operation, the link between the load and the lifting device is often a sling made of synthetic, alloy steel chain, or wire rope material. The way the sling is arranged to hold the load is called the rigging configuration, or hitch. Hitches can be made using just the sling or by using connecting hardware, as well. There are three basic types of hitches:

- Vertical
- Choker
- Basket

One of the most important parts of the rigger's job is making sure that the load is held securely. The type of hitch the rigger uses depends on the type of load to be lifted. Different hitches are used to secure, for example, a load of pipes, a load of concrete slabs, or a load of heavy machinery.

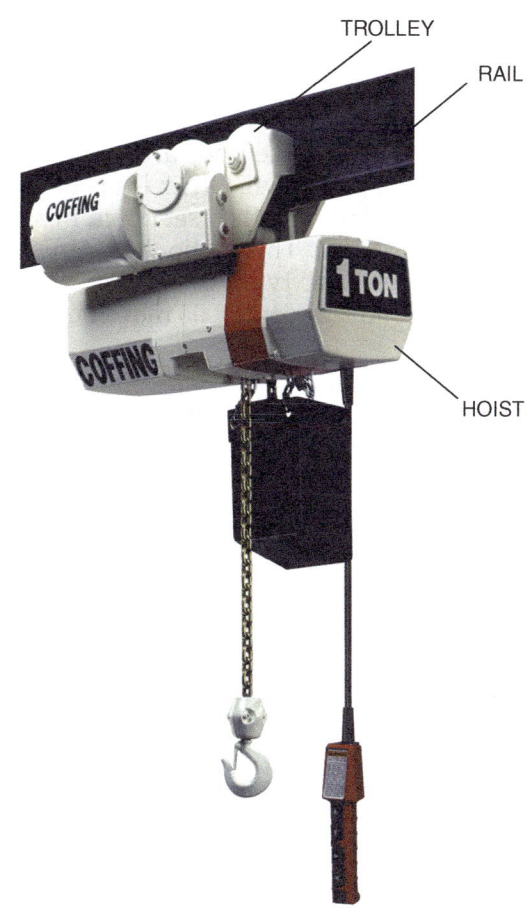

Figure 41 Hoist suspended from a trolley system.

RATCHET LEVER HOIST

Figure 42 Use ratchet lever hoists for vertical lifting.

00106-15 Introduction to Basic Rigging Module Six 23

Controlling the movement of the load once the lift is in progress is another important part of a rigger's job. Therefore, the rigger must also consider the intended movement of the load when choosing a hitch. For example, some loads are lifted straight up and then straight down. Other loads are lifted up, turned in midair, and then set down in a completely different place. This section examines how each of the three basic types of hitches is used to both secure a load and control its movement.

> **WARNING!**
> All rigging operations are dangerous, and extreme care must be used at all times. A straight up-and-down vertical lift is every bit as dangerous as a lift that involves rotating a load in midair and moving it to a different place. Only a qualified person may select the hitch to be used in any rigging operation.

1.5.1 Vertical Hitch

The single vertical hitch is used to lift a load straight up. This configuration forms a 90-degree angle between the hitch and the load. With this hitch, some type of attachment hardware, such as a shackle, is needed to connect the sling to the load (*Figure 43*). The single vertical hitch allows the load to rotate freely. To prevent the load from rotating, some method of load control, such as a tag line, must be used.

Another classification of hitch is the bridle hitch (*Figure 44*). The bridle hitch consists of two or more vertical hitches attached to the same hook, master link, or bull ring. The bridle hitch allows the slings to be connected to the same load without the use of such devices as a spreader beam, which is a stiff bar used when lifting large objects with a crane hook.

The multiple-leg bridle hitch (*Figure 45*) consists of three or four single hitches attached to the same hook, master link, or bull ring. Multiple-leg bridle hitches provide increased stability for the load being lifted. A multiple-leg bridle hitch is always considered to have only two of the legs supporting the majority of the load and the rest of the legs balancing it.

1.5.2 Choker Hitch

The choker hitch is used when a load has no attachment points or when the attachment points are not practical for lifting. The choker hitch is made by wrapping the sling around the load and passing it through one eye to form a constricting

Figure 43 Single vertical hitch.

loop around the load. In many configurations, the sling is wrapped around the load and passed through a shackle to form the constricting loop. In those situations, it is important that the shackle used in the choker hitch be oriented properly, as shown in *Figure 46*. Remember that the pressure of the load should never be placed on the pin of the shackle.

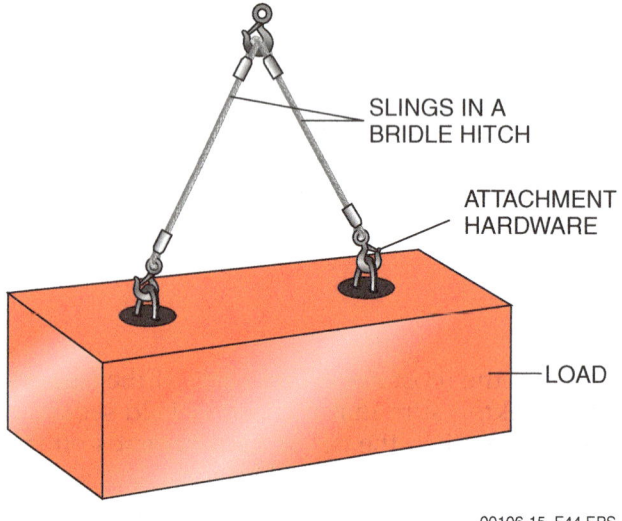

Figure 44 Bridle hitch.

24   NCCER – *Core Curriculum*   00106-15

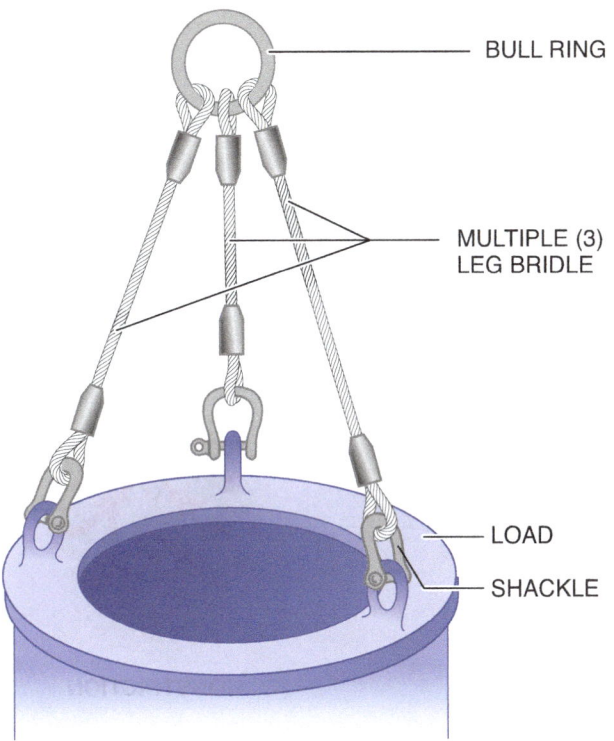

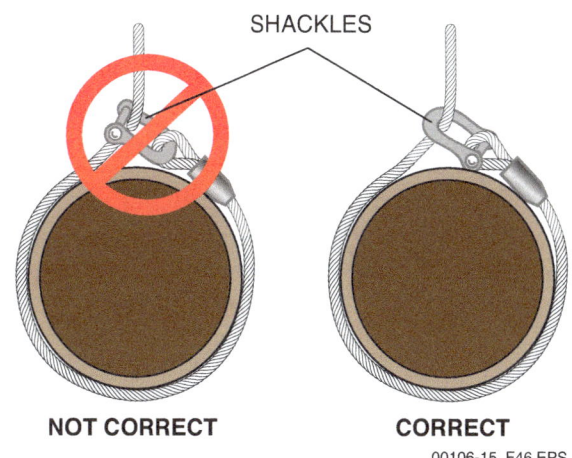

Figure 46 Choker hitches.

The choker hitch affects the capacity of the sling, reducing it by a minimum of 25 percent. This capacity reduction must be considered when choosing the proper sling. Further, the choker hitch does not grip the load securely. It is not recommended for loose bundles of materials because it tends to push loose items up and out of the choker. Many riggers use the choker hitch for bundles, mistakenly believing that forcing the choke down provides a tight grip. In fact, it serves only to drastically increase the stress on the choke leg (*Figure 47*).

When an item more than 12 feet (≈3.5 m) long is being rigged, the general rule is to use two choker hitches spaced far enough apart to provide the stability needed to transport the load (*Figure 48*).

To lift a bundle of loose items such as pipes and structural steel, or to maintain the load in a certain position during transport, double-wrap choker hitches (*Figure 49*) may be useful. A double-wrap choker hitch is made by wrapping the sling completely around the load, and then wrapping the choke end around again and connecting to the running end like a conventional choker hitch. This enables the load weight to produce a constricting action that binds the load into the middle of the hitch, holding it firmly in place throughout the lift.

Forcing the choke down will drastically increase the stress placed on the sling at the choke point. The double-wrap choker uses the load weight to provide the constricting force, so there is no need to force the sling down into a tighter choke; it serves no purpose. As with a single choker hitch, lifting a load longer than 12 feet (≈3.5 m) requires two double-wrap choker hitches.

Figure 45 Multiple-leg bridle hitch.

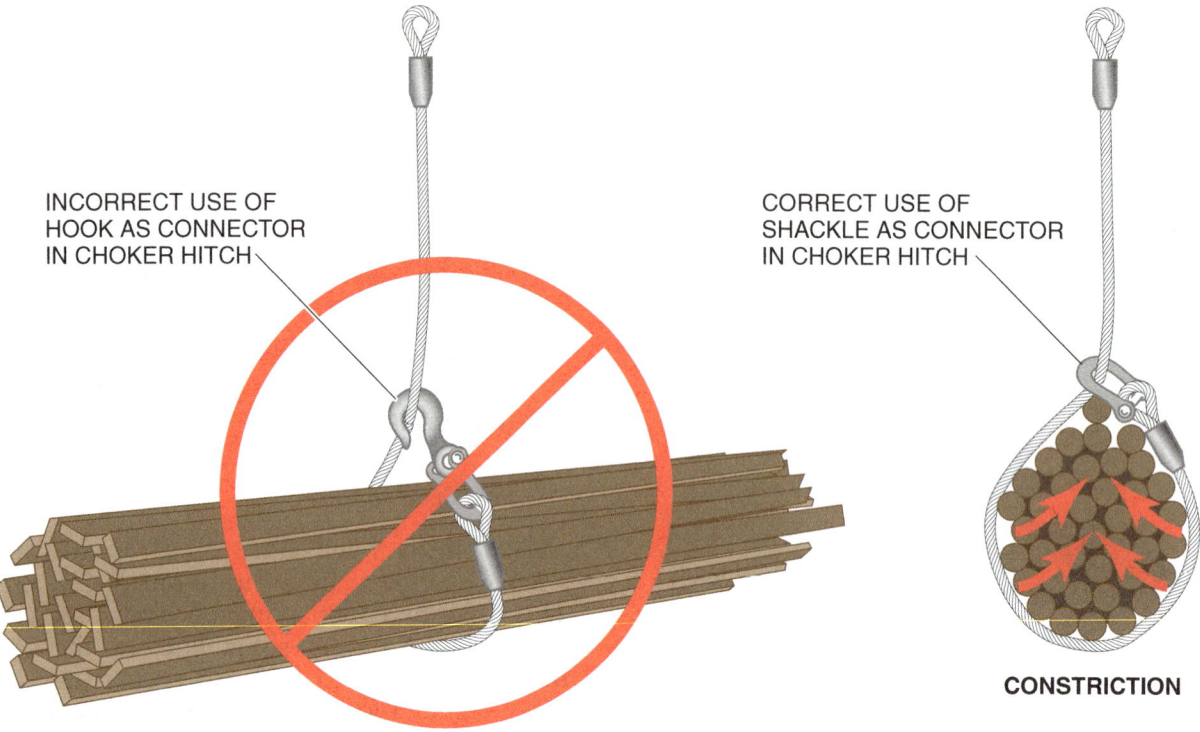

Figure 47 Choker hitch constriction.

Around the World

Crane Hand Signals

Hand signals of all types have different meanings around the world. In most cases, the issue only causes some confusion and embarrassment. However, when it comes to crane operations, incorrect hand signals can lead to disaster. Major lifting and rigging operations take place around the world, and not every member of the lift crew, including the crane operator, has received the same training or uses the same signals. On many projects, team members are not even from the same country.

A new set of hand signals has been published for adoption by the International Organization for Standardization (ISO) to make operations safer and more efficient worldwide. ISO 16715, entitled "Cranes—Hand Signals Used With Cranes" was developed by a subcommittee with international experts from countries such as Russia, Brazil, Australia, and the United States—31 countries in all. However, the objective was not to replace all unique signals worldwide. The real objective was to create a set of standardized signals used on job sites that involve multinational rigging and crane operation teams. Common operations in each country, involving local or regional workers, will likely continue to use the hand signals familiar to that region.

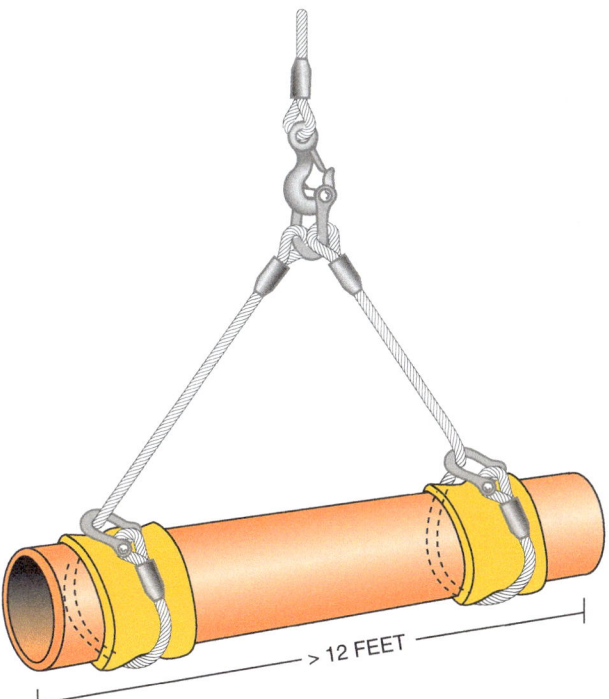

1.5.3 Basket Hitch

Basket hitches are very versatile and can be used to lift a variety of loads. A basket hitch is formed by passing the sling around the load (or, in some cases, through the load) and placing both eyes in the hook (*Figure 50*). Placing a sling into a basket hitch has the effect of doubling the capacity of the sling. This is because the basket hitch creates two sling legs from one sling.

A double-wrap basket hitch combines the constricting power of a double-wrap choker hitch with the capacity advantages of a basket hitch. This means it is able to hold a larger load more tightly. A double-wrap basket hitch requires a considerably longer sling length than a double-wrap choker hitch. If it is necessary to join two or more slings together, the load must be in contact with the sling body only, not with the hardware used to join the slings. The double-wrap basket hitch provides support around the load. As is true of the double-wrap choker hitch, the load weight provides the constricting force for the hitch.

> **CAUTION**
>
> A basket hitch should not be used to lift loose materials. Loads placed in a basket hitch should be balanced.

1.5.4 The Emergency Stop Signal

A major aspect of rigging safety involves maintaining clear communication between the crane operator and the designated signal person on the ground. This communication is normally accomplished using common signals—either verbal signals given by radio or hand signals. Hand signals used in rigging operations in the United States have been developed and standardized by the American Society of Mechanical Engineers (ASME) for all cranes. With the exception of the Emergency Stop signal, hand signals can only be given by the designated signal person on the ground.

Figure 48 Double choker hitches.

The Emergency Stop signal used in rigging operations is shown in *Figure 51*. In the event of an emergency, this signal can be given by anyone on the ground within sight of the crane operator. The Emergency Stop signal is made by extending both arms horizontally with palms down, and then quickly moving the arms back and forth by repeatedly extending and retracting them.

> **WARNING!**
> Although this module provides instruction and information about common rigging equipment and hitch configurations, it does not provide any level of certification. Any questions about rigging procedures or equipment should be directed to an instructor or supervisor. Always refer to the manufacturer's instructions for any type of rigging equipment.

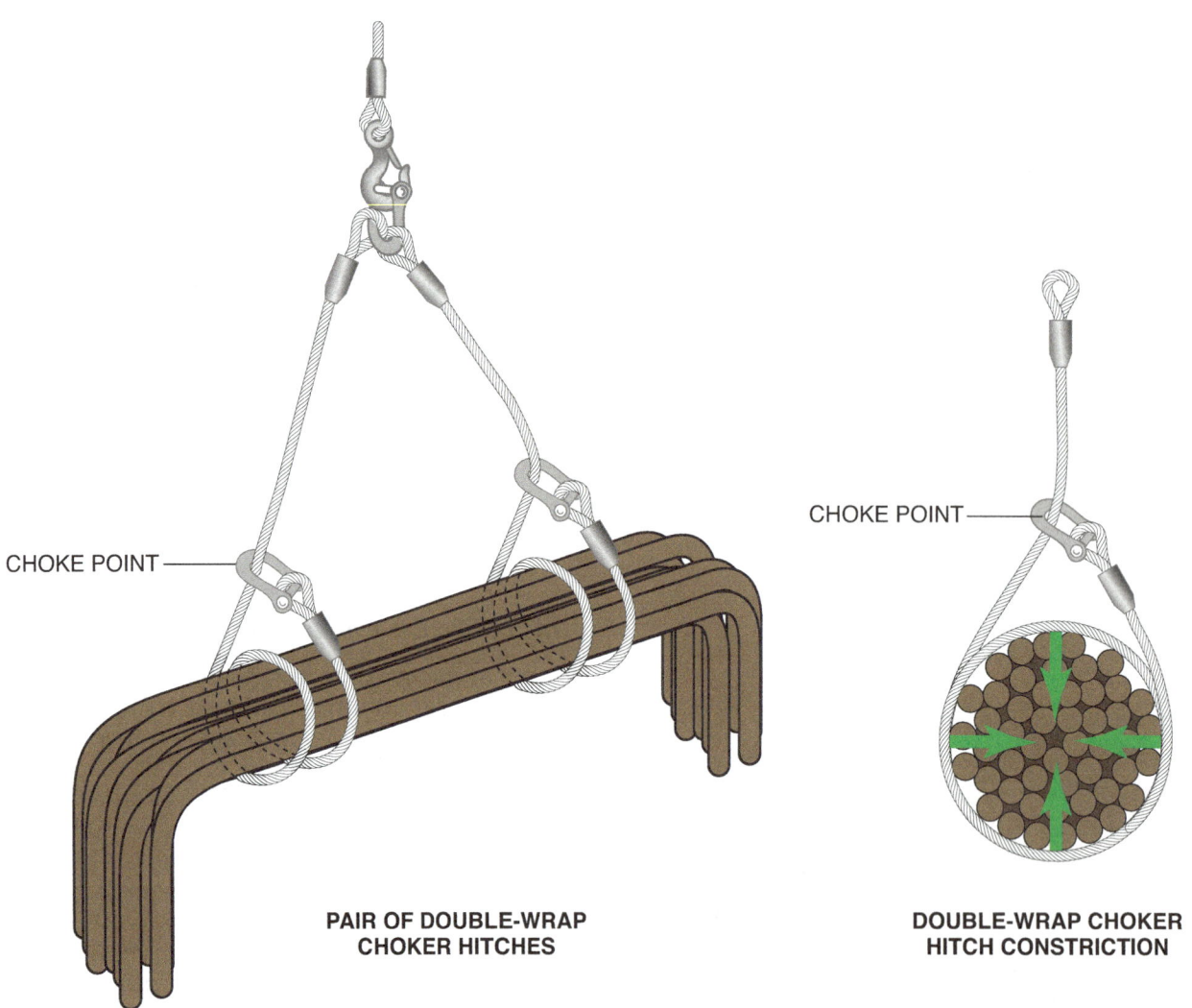

Figure 49 Double-wrap choker hitches.

Figure 50 Basket hitch.

Figure 51 Emergency Stop hand signal.

Additional Resources

Bob's Rigging and Crane Handbook, Bob De Benedictis. 2006. Leawood, KS: Pellow Engineering Services, Inc.

Mobile Crane Manual, Donald E. Dickie, D. H. Campbell. 1999. Toronto, Ontario, Canada: Construction Safety Association of Ontario.

Rigging Handbook, Jerry A. Klinke. 2012. Stevensville, MI: ACRA Enterprises, Inc.

Rigging Manual, 2005. Toronto, Ontario, Canada: Construction Safety Association of Ontario.

Rigging, James Headley. 2012. Sanford, FL: Crane Institute of America, Inc.

1.0.0 Section Review

1. Slings are commonly made out of _____
   a. synthetic fibers, cast-iron links, and wire rope
   b. fiberglass strands, exotic metals, and synthetic rope
   c. synthetic material, alloy steel, and wire rope
   d. carbon fiber, stainless steel, and twisted pair cable

2. If the identification tag on a synthetic sling is missing or illegible, the sling must be _____.
   a. re-tagged by a qualified rigger on site
   b. removed from service immediately
   c. used only for minimal loads
   d. tested on a load to determine its capacity

3. The size of a shackle is determined by the diameter of its _____.
   a. body
   b. shank
   c. eye
   d. pin

4. A type of rigging equipment that uses a pulley system to provide a mechanical advantage for lifting a load is a _____.
   a. fulcrum
   b. hitch
   c. hoist
   d. come-along

5. When a load has no attachment points or when the attachment points are not practical for lifting, it is best to use a _____.
   a. bridle hitch
   b. vertical hitch
   c. multiple-leg hitch
   d. choker hitch

SUMMARY

Although rigging operations are complex procedures that can present many dangers, a lift executed by fully trained and qualified rigging professionals can be a rewarding operation to watch or participate in. In order to accomplish a successful rigging operation, workers need to have a comprehensive understanding of the equipment used in rigging as well as some basic procedures for connecting and lifting loads. This module provided an overview of various types of rigging equipment and some common hitch configurations used for moving material and equipment from one location to another. Understanding these basic rigging standards will help provide the groundwork for a safe, productive, and rewarding construction career.

Review Questions

1. Identification tags for slings must include the _____.
   a. type of protective pads to use
   b. type of damage sustained during use
   c. color of the tattle-tail
   d. manufacturer's name or trademark

2. The type of wire rope core that is susceptible to heat damage at relatively low temperatures is the _____.
   a. fiber core
   b. strand core
   c. independent wire rope core
   d. metallic link supporting core

3. Synthetic slings must be inspected _____.
   a. once every month
   b. visually at the start of each work week
   c. before every use
   d. once wear or damage becomes apparent

4. An alloy steel chain sling must be removed from service if there is evidence that _____.
   a. the sling has been used in different hitch configurations
   b. replacement links have been used to repair the chain
   c. the sling has been used for more than one year
   d. strands in the supporting core have weakened

5. A piece of rigging hardware used to couple the end of a wire rope to eye fittings, hooks, or other connections is a(n) _____.
   a. eyebolt
   b. hitch
   c. shackle
   d. U-bolt

6. A lifting clamp is most likely to be used to move loads such as _____.
   a. steel plates
   b. piping bundles
   c. concrete blocks
   d. plastic tubing

7. Chain hoists are able to lift heavy loads by utilizing a _____.
   a. rope and pulley system
   b. rigger's strength
   c. stationary counterweight
   d. gear system

8. Before attempting to lift a load with a chain hoist, make sure that the _____.
   a. hoist is secured to a come-along
   b. load is properly balanced
   c. tag lines are properly anchored
   d. tackle is connected to its power source

9. A hitch configuration that allows slings to be connected to the same load without using a spreader beam is a _____.
   a. double-wrap hitch
   b. choker hitch
   c. bridle hitch
   d. basket hitch

10. To make the Emergency Stop signal that is used by riggers, extend both arms _____.
    a. horizontally with palms down and quickly move both arms back and forth
    b. directly in front and then move both arms up and down repeatedly
    c. vertically above the head and wave both arms back and forth
    d. horizontally with clenched fists and move both arms up and down

Trade Terms Quiz

Fill in the blank with the correct term that you learned from your study of this module.

1. A simple rope-and-pulley system called a(n) _____ is used to lift loads.
2. Use a single ring called a(n) _____ to attach multiple slings to a hoist hook.
3. The _____ is the distance between the master link of the sling to either the end fitting of the sling or the lowest point on the basket.
4. The way the sling is arranged to hold the load is called the rigging configuration, or _____.
5. A(n) _____ hitch uses two or more slings to connect a load to a single hoist hook.
6. A(n) _____ uses a pulley system to give you a mechanical advantage for lifting a load.
7. A(n) _____ is used to move loads such as steel plates or concrete panels without the use of slings.
8. The _____ is the total amount of what is being lifted.
9. The tension applied on the rigging by the weight of the suspended load is called the _____.
10. The _____ is the main connection fitting for chain slings.
11. To form an endless-loop web sling, _____ the ends together.
12. _____ equals the lengthwise distance it takes for one strand of wire to make one complete turn around the core.
13. Standard eye-and-eye slings have eyes on the same _____, whereas twisted eye-and-eye slings have eyes at right angles to each other.
14. The _____ is the link between the load and the lifting device.
15. The maximum load weight that a sling is designed to carry is called its _____.
16. Examples of _____ for synthetic slings include a missing identification tag, a puncture, and crushing.
17. Use a(n) _____ to attach a sling to a load.
18. Often found on a crane, a grooved pulley-wheel for changing the direction of a rope's pull is called a(n) _____.
19. The parts of the sling that reach from the attachment device around the object being lifted are called the _____.
20. A(n) _____ is a group of wires wound around a center core.
21. Riggers use a(n) _____ to limit the unwanted movement of the load when the crane begins moving.
22. If the _____ is showing, the sling is not safe for use.
23. A wire rope sling consists of high-strength steel wires formed into strands wrapped around a supporting _____.
24. An eyebolt is a piece of rigging hardware with a(n) _____, which means it has a series of spiral grooves cut into it.
25. An individual that has a college degree in Safety Technology would be considered a _____ in the field of safety.
26. To prevent the load from rotating freely, you must use some method of _____.
27. _____ slings are made of high-strength steel wires formed into strands wrapped around a core.
28. The maximum amount of weight a structure can safely support is the _____.

29. A(n) _____ is used to determine if an endless sling has been overloaded.

30. If a cable sling is found with the strands become untwisted in a section, the problem is referred to as _____.

31. A worker capable of identifying existing or potential hazards in the work area and authorized to take steps to eliminate such hazards is referred to as a(n) _____.

Trade Terms

Block and tackle
Bridle
Bull ring
Competent person
Core
Hitch
Hoist
Lifting clamp

Load
Load control
Load stress
Master link
One-rope lay
Plane
Qualified person
Rated capacity

Rejection criteria
Rigging hook
Shackle
Sheave
Sling
Sling legs
Sling reach
Splice

Strand
Tag line
Tattle-tail
Threaded shank
Unstranding
Warning yarn
Wire rope

Cornerstone of Craftsmanship

John Stronkowski
Industrial Management & Training Institute
Director of Education

How did you get started in the construction industry?
I worked in the trades to pay my way through college. At that time, it was a means to an end.

Who or what inspired you to enter the industry? Why?
My family members worked in various trades. My grandfather was a master carpenter and that served as an inspiration to me.

What do you enjoy most about your career?
I enjoy the teaching aspect the most. It allows me to share my knowledge and experience, and hopefully impact the lives of my students.

Why do you think training and education are important in construction?
Properly trained people in the trades are crucial to the safety and welfare of the general public. Mistakes and shortcuts have significant costs beyond money.

Why do you think credentials are important in construction?
Documented credentials represent a properly trained tradesperson who is up-to-date with today's technology.

How has training/construction impacted your life and career?
In my dual role as a multiple-licensed tradesperson and a Roman Catholic priest, the construction field has allowed me to be in charge of all major construction projects for the Diocese of Bridgeport, CT. With my vast knowledge of different yet related systems, I am able to provide added benefits to the church without extraordinary costs to our parishioners.

Would you recommend construction as a career to others? Why?
Yes, definitely. The trades offer highly rewarding careers. When you start a construction job and see it through to the end, it gives you a sense of accomplishment.

What does craftsmanship mean to you?
It is professionalism at its highest level and being proud of the work accomplished.

Trade Terms Introduced in This Module

Block and tackle: A simple rope-and-pulley system used to lift loads.

Bridle: A configuration using two or more slings to connect a load to a single hoist hook.

Bull ring: A single ring used to attach multiple slings to a hoist hook.

Competent person: An individual capable of identifying existing and predictable hazards in the surroundings and working conditions which are unsanitary, hazardous, or dangerous to employees; an individual who is authorized to take prompt corrective measures to eliminate these issues.

Core: Center support member of a wire rope around which the strands are laid.

Hitch: The rigging configuration by which a sling connects the load to the hoist hook. The three basic types of hitches are vertical, choker, and basket.

Hoist: A device that applies a mechanical force for lifting or lowering a load.

Lifting clamp: A device used to move loads such as steel plates or concrete panels without the use of slings.

Load: The total amount of what is being lifted, including all slings, hitches, and hardware.

Load control: The safe and efficient practice of load manipulation, using proper communication and handling techniques.

Load stress: The strain or tension applied on the rigging by the weight of the suspended load.

Master link: The main connection fitting for chain slings.

One-rope lay: The lengthwise distance it takes for one strand of a wire rope to make one complete turn around the core.

Plane: A surface in which a straight line joining two points lies wholly within that surface.

Qualified person: A person who, through the possession of a recognized degree, certificate, or professional standing, or one who has gained extensive knowledge, training, and experience, has successfully demonstrated his or her ability to solve problems relating to the subject matter, the work, or the project.

Rated capacity: The maximum load weight a sling or piece of hardware or equipment can hold or lift; also referred to as the working load limit (WLL).

Rejection criteria: Standards, rules, or tests on which a decision can be based to remove an object or device from service because it is no longer safe.

Rigging hook: An item of rigging hardware used to attach a sling to a load.

Shackle: Coupling device used in an appropriate lifting apparatus to connect the rope to eye fittings, hooks, or other connectors.

Sheave: A grooved pulley-wheel for changing the direction of a rope's pull; often found on a crane.

Sling: Wire rope, alloy steel chain, metal mesh fabric, synthetic rope, synthetic webbing, or jacketed synthetic continuous loop fibers made into forms, with or without end fittings, used to handle loads.

Sling legs: The parts of the sling that reach from the attachment device around the object being lifted.

Sling reach: A measure taken from the master link of the sling, where it bears weight, to either the end fitting of the sling or the lowest point on the basket.

Splice: To join together.

Strand: A group of wires wound, or laid, around a center wire, or core. Strands are laid around a supporting core to form a rope.

Tag line: Rope that runs from the load to the ground. Riggers hold on to tag lines to keep a load from swinging or spinning during the lift.

Tattle-tail: Cord attached to the strands of an endless loop sling. It protrudes from the jacket. A tattle-tail is used to determine if an endless sling has been stretched or overloaded.

Threaded shank: A connecting end of a fastener, such as a bolt, with a series of spiral grooves cut into it. The grooves are designed to mate with grooves cut into another object in order to join them together.

Unstranding: Describes wire rope strands that have become untwisted. This weakens the rope and makes it easier to break.

Warning yarn: A component of the sling that shows the rigger whether the sling has suffered too much damage to be used.

Wire rope: A rope made from steel wires that are formed into strands and then laid around a supporting core to form a complete rope; sometimes called cable.

Additional Resources

This module presents thorough resources for task training. The following resource material is suggested for further study.

Bob's Rigging and Crane Handbook, Bob De Benedictis. 2006. Leawood, KS: Pellow Engineering Services, Inc.

Mobile Crane Manual, Donald E. Dickie; D. H. Campbell. 1999. Toronto, Ontario, Canada: Construction Safety Association of Ontario.

Rigging Handbook, Jerry A. Klinke. 2012. Stevensville, MI: ACRA Enterprises, Inc.

Rigging Manual, 2005. Toronto, Ontario, Canada: Construction Safety Association of Ontario.

Rigging, James Headley. 2012. Sanford, FL: Crane Institute of America, Inc.

Figure Credits

Mammoet USA South, Inc., Module opener, Figure 2, SA01

Topaz Publications, Inc., Figure 1

Slingmax, Inc., Figure 4

Lift-All Company, Inc., Figures 5–7, 8A–8C, 9–11, 19, 22N–P, 24

Lift-It Manufacturing Co., Inc., Figures 12–15

Columbus McKinnon Corporation, Figures 16, 26, 29, 36A, 40

Ed Gloninger, Figures 22A-M

Courtesy of The Crosby Group LLC, Figures 27, 28, 31, 36B, 36C, 43

Volvo Construction Equipment, Figures 30, 45B (photo), 48B (photo)

J.C. Renfroe & Sons, Figures 33, 34

Gunnebo Johnson Corporation, Figures 36D, 36E

Walter Meier Manufacturing Americas, Figures 39, 42

Coffing Hoists, Figure 41

SME Brett Richardson, Starcon, Figure 50

Section Review Answer Key

Answer	Section Reference	Objective
Section One		
1. c	1.1.0	1a
2. b	1.2.1	1b
3. a	1.3.1	1c
4. c	1.4.0	1d
5. d	1.5.2	1e

NCCER CURRICULA — USER UPDATE

NCCER makes every effort to keep its textbooks up-to-date and free of technical errors. We appreciate your help in this process. If you find an error, a typographical mistake, or an inaccuracy in NCCER's curricula, please fill out this form (or a photocopy), or complete the online form at **www.nccer.org/olf**. Be sure to include the exact module ID number, page number, a detailed description, and your recommended correction. Your input will be brought to the attention of the Authoring Team. Thank you for your assistance.

Instructors – If you have an idea for improving this textbook, or have found that additional materials were necessary to teach this module effectively, please let us know so that we may present your suggestions to the Authoring Team.

NCCER Product Development and Revision
13614 Progress Blvd., Alachua, FL 32615

Email: curriculum@nccer.org
Online: www.nccer.org/olf

❏ Trainee Guide ❏ Lesson Plans ❏ Exam ❏ PowerPoints Other _____

Craft / Level: _____ Copyright Date: _____

Module ID Number / Title: _____

Section Number(s): _____

Description: _____

_____

_____

_____

_____

Recommended Correction: _____

_____

_____

_____

_____

Your Name: _____

Address: _____

_____

Email: _____ Phone: _____